COMMONSENSE
MEDICINE

Published by Lisa Hagan Books 2023

www.lisahaganbooks.com

ISBN: 978-1-945962-51-6

Cover photography: Frank McKenna http://blog.frankiefoto.com/

Interior layout by Simon Hartshorne

Cover art by Piper-Jayne Bozeman. Piper-Jayne is Dr. Jones's granddaughter. He wanted a cover showing how our profit oriented system has crippled the healing staff of Asclepius and led his wise serpent to eat its own tail. She did that with pizzazz.

COMMONSENSE MEDICINE

Healing from the Inside Out & Stopping the Next Pandemic

Dr. Lon Jones, DO

Contents

Setting the Stage ... 7
Dedication ... 9

Chapter One: From Ancient Greece to Modern Medicine: My Journey ... 11

Modern Hippocratic Medicine ... 19
Defense Medicine ... 25
The Genius of Simple Solutions ... 26
A Cosmic War ... 42
Let's Play Defense ... 45
Whispers of Asclepius ... 48
Tactics vs. Strategy ... 51
The Mind Virus ... 55

Chapter Two: Surviving the Next Pandemic ... 59

Chapter Three: Where this goes ... 67

And a little child shall lead them ... 72
A Paradigm of True Greatness ... 75
What Survival, and Surviving a Pandemic Really Means ... 79
Take the United States for example. ... 84
The Defensive Front Line ... 86
Commensal Biofilms to Invaders: 'No Vacancy' ... 97
Are Simple Defenses too Simple? ... 103
Curing Asthma ... 108
How We Got Here ... 111
Respiratory Defenses and the Pandemic ... 117

Chapter Four: Fixing Health Care 127

How We Got Here 134
Can We Get Some Empathy, Please 137
Survival of Healthcare 140
Your Health, Your Choice 149
Ending the Drug War to Make Us Healthier and Wealthier 154

We are at war with many things because wars are profitable. Among other things, they boost our GDP and pull us together behind a common cause. But our oldest and longest war is with microbes. Louis Pasteur (1822-1895) began this war and our brand of Western medicine when he showed us how the supposedly weak and defenseless microbes cause our diseases. We are still fighting, and it is exclusively an offensive war, but that's only one side, or a half, of the game.

Forgotten in our warfare is the life-long argument between Pasteur and Claude Bernard (1813-1878), the man honored as the father of modern physiology. Bernard argued that it is not so much the microbe that causes disease as it is the soil where it grows, i.e. us and our natural defenses. Both were right, but we have ignored the half that is us—Bernard's half that is made up of the defenses we all have that protect our bodies against the causes of illness and disease. Attack is more profitable, but the main reason we have ignored Bernard is the limits imposed on us by fear of the microbe. The only options for action when fear dominates are fight, flight, or freeze—thinking is constrained.

In a phrase repeated so often the source is unknown incoming medical students are told: half of what you'll learn in medical school will be shown to be either wrong or out of date within ten years, but no-one knows which half."

In this book we try to bring the halves together that make us and our communities whole. It is time to end the war and return to commonsense medicine!

Dedication

I was raised in the Mormon tradition, and with a brood of 14, I am a proud testament to our love of large families. My children are lawyers, engineers, professors, entrepreneurs, builders, and mothers and fathers. It was the first five of my children—all daughters—who opened my eyes to the patriarchal and tribal nature of the Mormon church. They certainly learned how to think for themselves, form their own opinions, and speak their truth. Now it's dad's turn to speak the truth he's learned from both studying history and practicing medicine for nearly 50 years.

I once asked a carpenter to build a table for my large family and when I told him how many—ten of us at the time—he proclaimed, "that's obscene." I retorted, "I'll take care of the quality, you can take care of the quantity," and ended up getting my table elsewhere. But I need even more room these days with the addition of two more children, two stepsons, and two foster children. It's to those children and their mothers that this book is dedicated.

CHAPTER ONE:

From Ancient Greece to Modern Medicine: My Journey

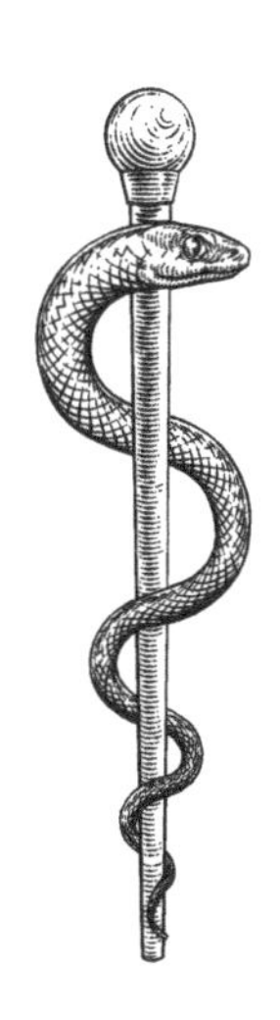

Like many great endeavors, the genesis of this book began with a need. My cute six-month-old granddaughter suffered from recurring ear infections, and her physician grandpa would do anything to help her—especially after my wife warned that such infections often lead to learning issues, and she would know because she teaches special ed. It put me on the road to inventing Xlear® (pronounced "clear"), a simple saline nasal spray with xylitol.

That treatment was certainly not without precedent. Finnish researchers have found that kids chewing xylitol sweetened gum five times a day in school had about half the problems with ear infections than those not chewing the gum. And the gum chewing was also not without precedent. Twenty years of research have shown xylitol to 'tame' the microbes that live on the plaque on our teeth that make the acid that eats through the enamel surfaces of our teeth to begin a cavity. Chewing the gum five times a day reduces decay by about 80%—it's a good idea. But the kids in my practice were too young to chew gum. The gum tames the decay causing microbes and the authors of the ear infection study said it too worked on the microbes causing ear infections—in fact they are cousins to the ones on our teeth. I put it in the back of the nose because that's where the microbes live that cause ear infections. The nose is a nidus, a nest of pests that cause lots of problems. From there they move to the sinuses, the middle ear, and downward to the lungs and bronchi.

But it wasn't always that way; it's a nidus because of environmental changes and wrongheaded therapeutical thinking. Fixing that is one purpose of this book.

In the beginning I knew what this spray did, but I really didn't know how it did it. In the intervening decades I have developed a pretty good idea of how. It's based mostly on the ideas of three researchers, Claude Bernard, J. William 'Bill' Costerton and Paul Ewald. They will be discussed in greater detail later, but each of them contributed to the idea that I call "defense medicine."

Every game we play has two parts: offense and defense. The offense wants to score and the defense wants to prevent that. It is much the same with our endless game with microbes. But there are some significant differences. We think it is good when our defense takes out the opposing quarterback, and we mistakenly see our use of antibiotics and their taking out of whatever is infecting us as equally good. We don't see the downside of how it stimulates antibiotic resistance that the World Health Organization predicts will be killing 10 million people annually by 2050. Defense medicine addresses the problem by changing the focus from killing to negotiating.

Bernard was the first to do so. In his long 19th Century argument with Pasteur on the cause of disease, Pasteur argued for the microbe. He gave us a weak, defenseless microbe as the cause and began the warfare mentality we continue today. But the growth of antibiotic resistant organisms shows the mistake in this approach—microbes are neither weak nor defenseless. Bernard argued that the cause was in the soil where the microbe was planted, which leads to looking at the defenses we have all obtained by millennia of playing this game. Those defenses are most robust at our openings where we are most vulnerable. Still, with our warfare mentality and offense focus, we have ignored most of them, even to the point of seeing them as illnesses and developing drugs to treat them. We have done this without seeing their defensive value or realizing that these FDA approved drugs cripple the evolutionary defenses and eliminate the survival value behind their expression in us. That's wrongheaded.

Bill Costerton introduced us to our friendly microbes that provide a large part of the defenses at our openings. It was his research that was behind our understanding of the problems of douching where vigorous washing removed helpful friendly bacteria and opened the door to pathogens. He also showed us how microbes are normally covered by an extensive matrix of sugar chains called glycans. Writing in 1977 he showed in his January 1978 *Scientific American* article "How Bacteria Stick" that it's the glycans on the microbes 'sticking' to the glycans covering our cells that begins an infection. He then explains three ways to interfere with this attachment and points out that such interference does not get into the microbe and so does not simulate the resistance that is so problematic with antibiotics.

His idea also found support from Paul Ewald in his book, *Evolution of Infectious Disease*, where he looks at the things we do to prevent contagion and shows how they impact microbial evolution. If we make it harder for microbes to spread, they tend to adapt in friendly ways. He shows that connection with a variety of interventions, from clean water and cholera to condoms and HIV, but they are all external. Nowhere does he talk about intervening inside us as Costerton describes, but both reach the same conclusion: if we can block microbial attachment to a host without hurting the microbe, they adapt in friendlier ways, and we win the game. That is what Defense medicine is doing; it's negotiating an end to our unwinnable war with microbes without threatening them—and nasal xylitol is the part dealing with our airway and its defenses.

My invention cleared up my granddaughter's ear infections and provided a royalty stream that enabled me and my wife to travel frequently. One of our trips took us to the Greek island of

Kos, where the father of Western medicine, Hippocrates, once lived and practiced medicine. These days he is known best for the Hippocratic Oath given to medical school graduates. It guides the conduct of physicians. Unfortunately, the medical system those graduates become part of may best be summed up more as "hypocritic" than Hippocratic. Hippocrates would tell us that we are hypocrites because we say we want what's best for our patients, but in practice we tend to serve ourselves and the landlords who own the healthcare system.

My wife and I visited Kos because of my curiosity about the ancient Greek physician who was as much a philosopher as a medical practitioner. He taught the teachers of medicine to impart the importance of prescribing only beneficial treatments, refraining from causing harm, and living an exemplary personal and professional life. Before going to medical school, I studied the history of science in graduate school and wanted to know more about the history of medicine. What better way to learn about Hippocrates than by walking the same ground he did thousands of years ago.

Hippocrates taught us that our food should be our drugs—an approach largely forgotten by Western medicine, which by its very definition excludes food as a form of medicine. Many foods have strong drug effects; we call them "Hippocratic drugs," as opposed to "FDA drugs." According to the National Cancer Institute, Western medicine is "a "system in which medical doctors and other healthcare professionals (such as nurses, pharmacists, and therapists) treat symptoms and diseases using drugs, radiation, or surgery."[1] So you mean to tell me that Western physicians take

1 *NCI Dictionary of Cancer terms.* National Cancer Institute. (n.d.). Retrieved May 17, 2022, from https://www.cancer.gov/publications/dictionaries/cancer-terms/def/western-medicine

an oath, go into practice, and immediately ignore one of the main teachings of the inspiration behind the oath? Yes, and that's just the tip of the iceberg, because Hippocrates' brand of medicine also gives control to the patient. That control is a critical part of being healthy and resilient, and removing food, that is under the patient's control, from the therapeutic options and only relying on FDA drugs contributes to the learned helplessness that plagues our system. Hippocrates also added to the patients' control in the quirky way of expecting them to have a dream about how they can be healed.

That approach doesn't fit where the Doctor knows best, and the patient passively accepts it as the word of God. Modern Western medicine does not give control, it takes it away. And rarely does it prescribe food as medicine.

But Ayurveda, a system of medical practice said to have originated more than 5,000 years ago in India, is steeped in the foundation of food as medicine. Hippocratic drugs, foods with drug effects, are used to treat symptoms and diseases, and in Western medicine we are learning how foods can keep us healthy and even help our drugs work better. The brand of medicine Hippocrates practiced is all about patient control, and the food we eat is within our control. He recommended fasting, too, which we also know is a healthy practice.

But there's another element to the medicine practiced by Hippocrates that's even more distant from today's health care. Sick people on Kos and throughout the ancient Mediterranean world went to physicians like Hippocrates, known as an Asclepiads—followers of Asclepius, the Greek deity of healing whose medical skill was said to be so great, he could raise the dead. Sick people who required extensive treatment visited an Asklepion: a temple of Asclepius dedicated to healing. Hundreds of them existed at various times throughout the Mediterranean.

Today we might call such a place a hospital, except the temples were more like sanitaria, offering good food, rest, healthy activities, and plenty of sunshine and fresh air. The Asklepion created an environment for healing and let nature do the rest, along with the skilled care provided by specially-trained healer-priests. And they were where Asclepius himself, by request, showed up as a visitor in dreams with messages of healing, a practice known as dream incubation. In addition to being a physician and a priest, an Asclepiad was also a dream interpreter. And the temples were *crawling* with snakes, and visitors slept on the ground among them.

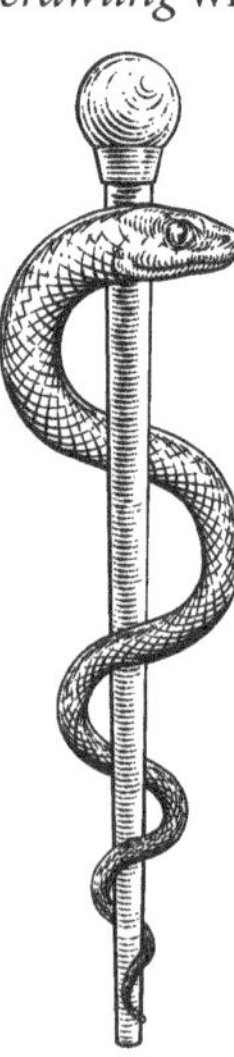

The Rod of Asclepius features a snake wrapped around a rod or staff. We can debate the symbolic meaning of the snake, but there's no debating the strong association between snakes and Asclepius. The snake is a symbol of healing because it's a creature that lives close to the earth and is highly sensitive to its environment, and it can transform by shedding its skin. The Staff of Asclepius with its snake is the symbol of the American Medical Association, and many other medical societies use it, too. Again, though, its origins and the brand of medicine it represents are all but lost.

Modern Hippocratic Medicine

The sort of medicine practiced by Hippocrates is a model of personal control and responsibility. The idea is largely lost in Western medicine, and it needs a comeback. Professor of medical sociology Aaron Antonovsky, in the seventies studied what makes people healthy and developed the concept of *salutogenesis,* meaning "the origins of health." It is based on the *sense of coherence* that he found in a cohort of healthy holocaust survivors, and coherence consists of the person having a sense of *control, understanding,* and *meaning.* Family therapists at the same time found essentially the same characteristics as the foundation of family resilience. To those with a sense of coherence, life is comprehensible, manageable, and meaningful. You are in control of your life and health, and those senses can be learned if we are willing. In this way, salutogenesis is an update on what Hippocrates lived and practiced thousands of years ago—and its application extends far beyond our personal health and family resilience (but that's in the second book of this series).

Psychologist Martin Seligman's experiments showed what happens to dogs when you remove control so they can't get out of an uncomfortable environment; he called it *learned helplessness.* Walk a hospital floor, especially one where serious disease is treated, and you will see all sorts of helplessness and the crushing of spirit that destroys health. The CIA liked Seligman's findings so much it put them to use in "enhanced interrogations" to "induce cooperation." In other words, torture produces learned helplessness, but you could also turn that statement around and say that learned helplessness is torture. It's certainly the loss of control, and the treatment is restoring that sense of control. Seligman didn't appreciate the negative press and recovered some when he showed service personnel suffering from combat trauma, which is essentially a form of learned helplessness, how to regain their *sense of control.*

Elinor Ostrom, a professor of political science, won the Nobel Prize for showing how the 'the tragedy of the commons'—the economic problem of overconsumption, underinvestment, and depletion of a common resource pool—could be rescued by sharing *control* of the commons with those who had been traditionally excluded from the decision-making. Those left out typically *understand* the problem, have the *meaning* gained by a shared community, but lack *control.* Restoring control builds the coherence that leads to a healthy common. In medicine, we call these left out people "patients."

In the same vein, researchers looking at fitness in ecosystems see it resulting from a diversity of elements and peaking with the interdependence of those elements. And that is true for ecosystems ranging from microbial to territorial—and I think it's true for social ecosystems like nation-states as well.

But building an interdependent social system is much more difficult because of paternalism. We have need to control our kids so they don't get into trouble, and it spreads to those we can control. But anyone with enough parenting experience can reach the same basic conclusion by observing the difference between giving and not giving their children a say in decisions that affect them. Psychologist Martin Seligman's experiments showed what happens to dogs when you remove control so they can't get out of an uncomfortable Before you can say "counterwill," you will see the resistance they put up when decisions are imposed on them. Counterwill is an instinct in children to resist what they perceive as unfair outside control, usually from a parent or authority figure. They resist coercion, and doing so helps them develop a sense of self. Child psychologists say that reaching a consensual decision with the child on the best path forward handles their counterwill because it shares *control.* And if you think about it, you will see that counterwill is not just in children; it seems to be a part of all living things. Personal control underpins the medicine practiced by Hippocrates. I call it commonsense medicine because it follows a rationale that even a child can understand. It not only improves personal health; moving control downward is also the key to the health of any living system, including our nation's. Somewhere along the way we lost track of Hippocrates and gave our control to experts and bureaucrats. As a physician, I can offer the same basic prescription for healing ourselves as I can for healing our nation, and it boils down to more control and coherence.

George Bernard Shaw, who received the Nobel Prize in Literature, saw the issue with Western medicine in terms of relinquishing control to the marketplace, as he wrote in *The Doctor's Dilemma*:

> That any nation . . . should go on to give a surgeon a [financial] interest in cutting off your leg, is enough to make one despair of political humanity.

Shaw recognized that the surgeon's interest, whether it be financial or ego-driven, leaned toward amputating a leg rather than trying to save it—after all, cutting off the leg, not saving it, is the surgeon's job. And if you think honor would stop a doctor from administering unnecessary medical procedures and treatments, you have probably never been in an operating room where the decisions are made. These are high-pressure environments where the circumstances are rarely black and white.

Take, for instance, what I learned while attending a continuing education program that looked at arthroscopic surgery. Arthroscopy is a common procedure for injured knees, and insurance pays minimally for it, but if a surgeon finds something in the knee that can be repaired with a snip or two, payment is substantially increased, and the conclusion of the presentation showed a preponderance of snips. A snip or two is not the same as amputation, but the principle is the same. When health care is a commodity, financial interests drive the decision-making.

But perhaps there is no example more blatant than the one provided by Dr. Sara "Joe" Baker, a pioneering physician and public health expert. In the early 1900s, during an epidemic of cholera and typhoid fever in New York, she instituted a hand-washing program in the city's schools. She writes in her autobiography, *Fighting for Life,* that it was so successful that a group of Brooklyn physicians asked the mayor to terminate the program because the lack of sick children hurt their practices![2]

Baker faced similar pushback from the medical aristocracy when she supported the Sheppard-Towner Act, a bill to create a nationwide program providing maternal and infant care. It built atop the New York City programs she instituted that led to steep declines in infant mortality. During the debate in Congress, a physician representing the AMA stated: "We oppose this bill because, if you are going to save the lives of all these women and children at the public expense, what inducement will there be for young men to study medicine?" Baker recalls that the committee

2 S. Josephine Baker, M.D.. *Fighting for Life.* 1940 Robert Hale Ltd. London. P. 157.

chairperson, Senator Morris Sheppard stiffened, leaned forward, and said pointedly, "Perhaps I didn't understand you correctly. You surely don't mean that you want women and children to die unnecessarily or live in constant danger of sickness so there will be something for young physicians to do?" The doctor from the AMA is said to have answered, "Why not? That's the will of God, isn't it?"[3]

Baker attributed the shortsightedness—some might say the criminality—to the emphasis on cure over prevention in the medical profession. Sound familiar? Not much has changed during the intervening century.

The Sheppard-Towner Act became law in 1921, and a century later, hand-washing and maternal and infant care are no longer opposed by the AMA, but the pecuniary interest to "look out for number-one" is still widespread in the medical profession and entrenched in the healthcare industry.

My own experience bears evidence. The Xlear nasal spray I created optimizes our nasal defenses, and, as a nice side effect, prevents asthma, too. As I gathered evidence to support my findings, my hospital-clinic closed and I went back to working ERs where studies like I was planning can't be done. Hoping to continue the research, I contacted several asthma researchers and tried to get their interest in continuing the study of the simple and potentially game-changing treatment of asthma. No one took me up on the offer. They didn't criticize my suggestion or offer suggestions, they just ignored me. When I asked my local physician's group I got the same response, but the wives talked and told mine they

---

3 Morantz-Sanchez, R. M. (2000). In Sympathy and science: Women physicians in American Medicine (pg. 302). essay, University of North Carolina Press.

were afraid of what preventing asthma would do to their income. It's also easy to see that the researchers' income and livelihood is based on studying the illness, not curing it.

'The money isn't in the cure, it's in the treatment,' as Chris Rock so memorably said in his criticism of HIV research, and that's the definition of hypocritic (for-profit) medicine, whereas Hippocratic medicine seeks the cure. Which do you prefer?

I prefer the one where people control their health, not where experts remove the control—even if they are doctors.

Once upon a time I was part of that crowd but never in the fullest sense—my experience studying the history of science before going into medicine impacted my view of medical practice and made me an outlier in the profession. I knew, for instance, about the life-long argument between Louis Pasteur and Claude Bernard about the cause of disease. Pasteur argued that the dastardly microbes are at fault. He started a war against them, enmeshing it in our culture and blinding us to other options. Bernard argued that the cause of disease is in the soil where the microbe is planted: our bodies, mostly via one of our more vulnerable openings where they had to cope with our physiologic defenses gained from years of evolutionary adaptation. Rather than pursuing a war against the microbial invaders by attacking them in their homeland, our first line of defense should be to increase the body's ability to reject the microbes before they can cause any damage at all, which brings us back to Defense Medicine.

Defense Medicine

When a microbe senses a threat, it leads to increased mutation to deal with the problem, and with billions of microbes looking for a solution it is invariably found. Nietzsche's statement, "what doesn't kill me makes me stronger," is very true for a community of microbes. That is the situation we are in today after more than a century relying on offense being the best defense. Even the drugs we use to defend ourselves are more often than not agents designed to kill the microbes.

Defenses work differently and defense medicine is the solution to our problem of antimicrobial resistance that the World Health Organization says will kill ten million people annually by 2050 if we don't find a way out.

Defense medicine is what xylitol does to the microbes causing tooth decay. It's message is 'shape up or ship out,' but it's delivered in a way that's not a threat.

Bill Costerton described that way in 1977 and explained to us how our friendly biofilm bacteria help make it happen. The plaque on our teeth is helpful if they don't contain the microbes causing tooth decay. Another helpful biofilm is in the female vagina. It's made up mostly of *Lactobacillus* species, microbes that make hydrogen peroxide and maintain an acidic environment that impedes other microbe from hanging on. A problem he saw, and was singularly influential at stopping, was the practice of douching. Vigorous washing can remove part of the biofilm, cripple its defensive role, and open the door to more infections. Using a neti pot to clean the nose does the same thing. Costerton summed it up in his article:

> One attractive aspect of an antibiotic directed against the glycocalyx is that it need not enter the host cells or the bacterial cells. thereby avoiding two common problems in antibiotic therapy: toxicity to host cells and the induction of bacterial resistance based on changes in the permeability of the bacterial-cell membrane.

Most researchers don't study the glycans because they are sugars that cannot be controlled, patented, and made into profitable drugs. Glycans are a very good example of Hippocratic drugs and we will discuss them more, much more, later. The point here is that our warfare with microbes can be significantly modified, if not resolved, simply by shifting to this non-threatening means of coping with microbes. We can in this way address the problem of the annual toll of ten million lives by 2050.

The Genius of Simple Solutions

I began my medical practice at the same time we were learning about the benefits of Oral Rehydration Therapy (ORT) for treating patients with cholera. Cholera kills not because of the infection but because our immune system sees it and its toxins as more dangerous than it really is and opens all the taps to wash it out. Victims die of dehydration caused by the body's overzealous defenses, not because of the microbe. ORT is a solution of salt, sugar, and water. Proportioned correctly, it activates a water pump in the GI tract. It's like drinking an IV. Most of my colleagues see ORT as an interesting treatment but would rather administer an IV. IVs are under their control, not the patient's, and the return (profitability) is orders of magnitude greater. On the other hand,

I see ORT as a commonsense treatment that optimizes gastroenteritis, which is not a medical condition needing treatment; it is the body's natural washing defense to remove pollutants from the GI tract.Which brings us back to my inspiration for inventing Xlear, the nasal spray that subsequently cured my granddaughter's recurrent ear infections. Jerry, my wife who avoided the onslaught of prescribed teaching by moving to special education, recognized years earlier that her students with language learning problems all had chronic ear infections with tubes in early childhood. The common medical response at the time was a procedure to place small tubes in their eardrums when they were babies to deal with the infections. It was a typical treatment that did nothing for the infections, it just made them so they didn't hurt, and it's not the hurt that interferes with learning, it's the infection in the middle ear. "If someone really cared about children," she said to me, "they would find a way to prevent ear infections."

Challenge accepted—caring and helping are two big reasons why I went into medicine, and my soft spot for kids is a big reason why I chose family practice. Jerome "Jerry" Klein, the coauthor of the standard medical textbook *Otitis Media in Infants and Children*, honored my wife's observations when he told us over dinner, "You tell your people that *this* Jerry agrees with *your* Jerry." I could not honor her more. Their agreement is about the connection between ear infections and learning problems. The two go together because the ear responds to infection with a washing defense similar to the defenses of the GI tract, secreting fluid to wash it out. But those infections are in the middle ear, and the Eustachian canal that connects the middle ear with the back of the nose, where the microbes begin, is often swollen so it has a hard time draining. The fluid pushes on the eardrum and it hurts–so tubes eliminate the pressure and the pain and allow some draining, but not enough. With recurrent infections the fluid tends to hang around and thicken—the English even call it "glue ear."

This happens in the middle ear, the home to the small bones that transmit the vibrations of the eardrum to the cochlea. The cochlea changes the vibrations to electrical impulses the brain recognizes as sound. Bones stuck in the glue don't move, and consequently the children can't hear. When ear infections are chronic, children can easily miss the developmental window for learning how sounds translate to words. That window opens early in the third trimester and closes at about two years of age. By then, the window for early childhood ear infections usually closes, too, and those children with the problem end up in special education classes with educators like my wife.

My paternal instincts kicked in and I was determined to give my granddaughter her best chance at life. I thought of all I'd learned from Hippocrates, Bernard, and others about playing defense instead of offense by aiding the body's natural defenses. I had earlier tried using saline nasal sprays for upper respiratory conditions and found them only marginally helpful. But shortly after Jerry's challenge I read about the Finnish chewing gum study.[4] Chewing the gum is standard practice in Finnish schools to prevent tooth decay.

The ear infection study's authors said that xylitol eliminates the bacteria—*Streptococcus mutans*—that cause tooth decay and does the same thing to their 'cousins'—*Streptococcus pneumoniae*—that live in the back of the nose and cause ear infections.

---

4 Uhari M, Kontiokari T, et al., "Xylitol Chewing Gum in Prevention of Otitis Media," *British Medical Journal*, November 9, 1996; 3137066): 1180–84.

Sweet serendipity—a solution fell into my lap, but our granddaughter was too young to chew gum, so we developed a nasal spray to deliver the xylitol in the back of her nose, where the germs live, prior to every diaper change.

Xylitol is common in many fruits and vegetables, and it turned out to be one of those foods that's also a very good medicine—it is a Hippocratic drug. I calculated that a person could use a xylitol saline spray every hour, 24 hours a day, both sides of the nose, and get about half a plum's worth of xylitol, and since it's not absorbed in the nasal passages it ends up in the stomach, just like when you eat a plum. I determined, in other words, that it was safe for our granddaughter.

I had my hospital pharmacy mix up the spray and directed her parents to use it with every diaper change, and her ear infections disappeared. Seeing the amazing benefits, her father suggested that I patent it, so I obtained a 'use patent'. Then I put my innovation to work with children who had chronic ear infections, tracking ten of them during the following year, and finding that the frequency of ear complaints plummeted by 95 percent.

That's how Xlear nasal spray was born. It's spelled with an X for xylitol, but X is pronounced the Finnish way, like a soft K, so it sounds just like "clear," which is what it does for your nose.

After seeing Xlear's success in preventing a variety of upper respiratory problems, I thought the FDA might be interested and filled out an Investigational New Drug application. The expense of that process was too much to handle personally, so I knocked on the pharmaceutical industry's door. Initially, they showed interest, with one of them even getting approval for a patent just like mine, but they turned their tail and ran away after finding out that xylitol is the active ingredient. That response is understandable;

in order to pay for the FDA demanded studies they need a "drug" they can control, patent, and make profitably. That's not possible with a food. If they pursued it and charged enough to pay for the studies people would just make it themselves.

So we paid attention to the legal advice to make a nose wash. But Xlear, while advertised as a nose wash, does not wash the nose. There are many products that do, but most of them wash everything from the nose including the mucus with all its defensive materials. It's like douching, and we no longer recommend that practice for the same reason–it washes out the defenses. What Xlear does instead is optimize our own defenses. More on that later.

Technically xylitol is a sugar alcohol, but it is neither a sugar nor an alcohol in the common sense of those words. Sugars are a family of chemicals with six carbon atoms in a fixed shape, with names like glucose, fructose, mannose, galactose, and fucose. They all have the same atoms but differ in structure and characteristics, and the glycan chains on our proteins and microbes are made up of these sugars and their simple modifications. Sugar alcohols differ from sugars by having one more hydrogen atom, which makes them flexible; and most importantly, they trick microbes looking for a meal or a place to take hold in the body. Sugar alcohols are champions at adapting to look appealing to bacteria, and that's what makes xylitol so effective at not only clearing infections but also preventing them. Bacteria grab hold of the decoy and never make it into the body.

This idea opens the doors to a couple of concepts that have not made it into medical education—and that's costly in both expense and lives. The first concept is biofilms and their infections. When a bacteria finds a place to hold on to us it grows to form a colony. When the microbes living in that colony sense a threat—lack of food, or an antibiotic—they release a messenger molecule and when the level of that messenger gets high enough the colony shifts and many of the microbes self sacrifice to build a matrix around the colony that protects them from the threat. That's then a biofilm. The process is called quorum sensing—it's like voting and we will talk about it more later. The point is that antibiotics don't penetrate their safe houses. Many of the recurrent ear infections I dealt with were from biofilms. Typically you prescribe an antibiotic and the infection subsides as the bacteria retreat to their safe house, only to recur when the antibiotic treatment ends. Such infections are also common on devices that are placed in the body, like tear duct inserts to treat dry eyes, joint replacements, stimulating devices, or tubes penetrating the skin used in surgical procedures that remain for too long. Antibiotics are seldom effective because of the safe house biofilms; the devices need to be removed.

Bill Costerton's 1978 article explains a bit about the sugars that make up what he called the glycocalyx. The first thing he introduced is the fact that microbes in nature are covered with a shield of sugars while the ones studied in the laboratories are not so covered. That's because we feed them in the lab so they don't need to build the covering. He called the cover a glycocalyx; it's made up of long chains of specific sugars and their complexes that are called glycans. These sugars are important, but we don't know very much about them. This from the *Essentials of Glycobiology* gives an idea:[5]

5 Varki A, Esko JD, Colley KJ. Cellular Organization of Glycosylation. In: Varki A, Cummings RD, Esko JD, et al., editors. Essentials of Glycobiology. 2nd edition. Cold Spring Harbor (NY): Cold Spring Harbor Laboratory Press; 2009. Chapter 3. Available from: https://www.ncbi.nlm.nih.gov/books/NBK1926/

> It is a remarkable fact that every free-living cell and every cell type within multicellular organisms is covered with a dense and complex layer of glycans. Even enveloped viruses that bud from surfaces of infected cells carry with them the glycosylation patterns of the host cell. [remember this!] Additionally, most secreted molecules are glycosylated and the extracellular matrices of multicellular organisms are rich in glycans and glycoconjugates. The matrices secreted by unicellular organisms when they congregate . . . also contain glycans. The reason for the apparent universality of cell-surface and secreted glycosylation is not clear, but it suggests that evolution has repeatedly selected for glycans as being the most diverse and flexible molecules to position at the interface between cells and the extracellular milieu.

Glycans are essentially everywhere on living cells, thousands of them extend outward from an average cell where they interact with what's outside, but our researchers have mostly ignored them. They are both too small for many to study and not profitable enough to pay for the research. The proteins that you read and hear about in the media are never just proteins, they're always glycoproteins. And it's these 'glycans', on both the microbes, our cells, and the proteins, that do the binding. Costerton pointed out the potential benefit of addressing this binding as a simple and effective way of coping with infection. In his words:[6]

6 Costerton JW, Geesey GG, Cheng KJ. How bacteria stick. Sci Am. 1978 Jan;238(1):86-95. doi: 10.1038/scientificamerican0178-86. PMID: 635520.

If adhesion has a central role in the success of pathogenic bacteria. then the prevention of adhesion should be an effective way to prevent or combat bacterial infection. It should be possible to develop a new class of antibiotics that interfere with glycocalyx formation or function in specific pathogens. There are at least three ways in which such inhibition might be achieved. One way would be to disrupt the synthesis of glycocalyx fibers. The bacterial polymerase that links sugar molecules to form these fibers should be inhibited if it is presented with a compound that mimics its normal substrate and therefore occupies the enzyme's active site. but that cannot be processed to build the normal polysaccharide fiber. In the absence of such fibers there would be no glycocalyx. no adhesion and no resistance to white cells. One might also find a compound that would occupy and block the active site of a lectin mediating the adhesion of bacterial glycocalyx fibers to the fibers of host cells.

Finally. it should be possible to block the "receptor" sites on host cells, that is, the glycoprotein fibers to which bacterial fibers adhere directly. One attractive aspect of an antibiotic directed against the glycocalyx is that it need not enter the host cells or the bacterial cells. thereby avoiding two common problems in antibiotic therapy: toxicity to host cells and the induction of bacterial resistance based on changes in the permeability of the bacterial-cell membrane.

To restate his suggestions there are three ways to block adherence:

1. Interfere with the formation of the glycan chains.
2. Interfere with the attachment of the chain to the protein on the cell surface.
3. Interfere by competing at the binding site.

Three simple approaches that, as Costerton points out, don't threaten the microbe and don't lead to the problem we have today with antimicrobial resistant organisms. So why hasn't our health care industry pursued any of these paths? Could it be that glycans are natural sugars and simple compounds that cannot be patented and made into profitable drugs?

Interestingly xylitol fits all three of Costerton's suggestions. The first requires the microbe to ingest the xylitol because glycan formation is done in the mitochondria inside the cell. Uhari and his group in Finland, the ones who did the chewing gum-ear infection study, found that the microbes ate the xylitol but, normally living off the six carbon sugars plentiful in animals, they lacked the enzymes to digest five carbon sugars like xylitol. So xylitol is available inside the cell to block the process. And it's even key in it, to address Costerton's second suggestion, since xylose, the parent sugar molecule of xylitol, is commonly the glycan that binds with the cell. The third option is Costerton's favored; it's competing at the binding site, and it's the simplest explanation for the findings behind the lab studies showing xylitol's ability to remove many microbial pathogens from their preferred binding sites. This includes not only the ones causing ear, sinus and lung infections, but the virus causing COVID., and many more.

Xylitol is commonly used as a sugar substitute; it looks like sugar, tastes like sugar, can be used like sugar, and because it is a five-carbon monosaccharide it has only about one-third the calories. And it cures diabetes in experimental diabetic rats by replacing sugar in their diets. But if you are not used to it the body doesn't absorb it well, so it can cause diarrhea—or just keep you regular. Sorbitol works the same way and is commonly used in nursing homes for this purpose. You can buy xylitol at some grocery stores, though health food stores are more likely to stock it because it costs more than the cheap, heavily subsidized, sugar we find on grocery shelves.

Xylitol can't be controlled or patented or made into an FDA-approved drug. If something can't be made into a drug, the drug industry wants nothing to do with it. They deal with drugs they can control. With the cost of FDA approval being so prohibitive, they need control so they can make a profit. It's a reason why they charge jacked-up drug prices—they need to pay for all the advertising they are allowed by the FDA to do—but if the "drug" is available at your grocery store, what's the point? People can make it themselves.

My patent is a use patent, meaning that no one can sell a product for spraying xylitol into the nose, but nothing's stopping you from making it yourself. Do what I did—add a teaspoon of xylitol to an ounce and a half bottle of saline nasal spray. Xlear is not an FDA drug; it's a commonsense, Hippocratic drug. It's like soap for the nose but without the burn or the bubbles—and it doesn't clean, it helps our own cleaning processes work better.

Early on, at the stage where I knew *what* xylitol in a nasal spray does, but was still blind to *how* it did it, I filed an IND (Investigational New Drug) application with the FDA. It didn't take long to see it wasn't going anywhere. The downside of not being FDA-approved is the manufacturer cannot make claims about its medical benefits, and during the COVID-19 pandemic it sure would have been nice to tell the public about the benefits of a clean airway—and my son did.

The company Jerry and I founded was copied by my entrepreneurial son, who left the oil rig business after a few close brushes with death, which helped my willingness to allow it. He took over making the spray so that I could continue practicing medicine. At the beginning of the COVID pandemic the company supported research that found the great effectiveness of xylitol and other sugar alcohols like sorbitol and erythritol for coping with COVID, and that the preservative they used, grapefruit seed extract, was toxic to the virus. If you want to concentrate on dealing with the virus in an offensive manner there are online recipes for making this extract, but it also feeds resistance. Most likely the xylitol works by preventing the attachment of the virus to the binding sites in our upper airway—Costerton's third avenue. The responsible thing to do with this crucial information for fighting the pandemic was to let people know about it, but regulatory agencies saw it differently—those are drug claims, only to be made by FDA approved drugs. He did it anyway.

Removing this type of information from the public's awareness is wrongheaded. People need to feel in control of their health. Providing them with a simple way of responding to a catastrophic emergency would have gone far in alleviating some of the fear as we wrestled with a viral mutation that was said to be on par with the Spanish Flu. Instead, in a mad rush, we gave control to the experts in their high-tech labs, who rolled out complex and costly new technologies at Warp speed that many see as insufficiently tested—but that's OK because the government has assumed responsibility for ill effects—if you can convince the government that the side effects rally belong to the insufficiently studied FDA 'emergently' approved drug.

Robbing the patient of control over their health and giving that control to the doctors in charge is a major problem with Western medicine. Even when Asclepius is the doctor giving the medical advice, it's still up to you to interpret and implement it. Whether right now or during the times of ancient Greece, the 'Thomas Rule', also referred to as the Thomas theorem, is still in effect, which sprang from William and Dorothy Thomas's sociological insights from a century ago: "If men define situations as real, they are real in their consequences."[7] In other words, the perception of health can be a self-fulfilling prophecy.

I once listened to a hospital chaplain tell the story of an ICU patient who was expected to die that night. The nurses asked the chaplain to perform the last rites. He did and told the man his sins were forgiven. Next morning, the patient woke up, told everyone he was healed, and that he wanted to go home. Believing he was healed, his body responded in kind—the ultimate "real consequence."

Doesn't matter who you call God or what you think it is, or whether belief is a placebo or the activator of a healing force we don't yet understand. The medical literature is full of astounding cases of spontaneous remissions and healings, and outside of the relatively few doctors who admit that something larger must be at work and want to know what it is. Even those looking at the remains of the crutches at the ancient Asclepia agree that something worked. But the reaction in the medical community is to

7 Thomas, William I., and Dorothy S. Thomas. 1929. The Child in America (2nd ed.). Alfred A. Knopf. p. 572.

ignore such healings or try to explain them away, missing out on a gold mine of insight into how the body heals.

Death is also a sort of healing. It's what Socrates meant with his last words, spoken to his friend Crito as he waited for the hemlock to kick in: "Crito, we owe a cock to Asclepius; pay it and don't forget." The sacrifice of a cock was the usual offering as thanks for delivering healing. Socrates believed his death to be the cure for the disease of life, a point of view that may have been influenced by the fact that he'd been sentenced to death for "impiety" and corruption of the youth. In other words, he spoke the truth as he knew it and refused to shut up when it rubbed certain powerful people the wrong way.

I owe a debt to Asclepius, too, for the specific instructions that came to me. *"Put 5 grams of xylitol in a 45 ml bottle of Ocean* [saline nasal spray]". It wasn't the first time I'd received insights while dreaming or while lying in bed just after waking up, but in this case, it changed the direction of my life and gave me the treatment for my granddaughter's ear infections that I so desired.

Insights that come in altered states of consciousness like dreaming are credited for many great inventions and discoveries. Some prolific dreamers like Albert Einstein call it "mental play" and don't explicitly credit supernatural sources for the discoveries delivered to them in their dreams, but others like Srinivasa Ramanujan do. His case is astonishing. One of the all-time great mathematicians, he learned mathematics from a textbook he scrounged up as a teenager living in extreme poverty in India, then proceeded to blow the minds of the best-known mathematicians of his time. "A magical genius" they said of him. A normal genius can explain their insights to others and more or less make people understand how they were reached. Einstein faced some difficulty getting his theory of relativity accepted by the physicists of his time (though these days you can get a good laugh by reading *A Hundred Authors Against Einstein* and the reasons why many prominent academics thought they were right, and he was wrong). Quantum physics is a bit above that—relatively few people truly understand it—but a magical genius is an entirely different ballpark—no one understands how it works, but it does. Ramanujan related that the goddess Namagiri visited him in his dreams and gave him the formulae he wrote down and later shared.

Like the visitors who sought healing in the temples of Asclepius, Ramanujan asked for visitations in his dreams and apparently received them. Mathematicians who study his writings find it difficult to reconcile his complete lack of understanding of some of the most basic concepts in advanced mathematics with the magical genius he often displayed. He was a self-trained peasant, but his natural intelligence was off the charts, and it is no surprise that his dreams reflected his brilliance. Early on many (some say a third) of his theorems were questioned, but over the years, with others proving them, his rate is in the 90% range. The truly inspired mathematical concepts he delivered went so far beyond his mediocre education, all rational explanations for his genius simply fall short. In my opinion, the credit can only be given to a supernatural force of inspiration that he himself credited: his intellectual patroness, Namagiri.

Over the years, my dreams, too, have been full of whispers from something I cannot explain rationally. I am both honored and humbled by the insights provided, though I admit that I do not know whether they come from the gods, the "mental play" behind Einstein's discoveries, or just an unstudied aspect of the human brain, but I know they are important. They need attention, and they need verification because they all suffer from the believer's bias. Ramanujan could not demonstrate the derivations of some of his formulae, and it's important to do so because his trust in his source, Namagiri, was so great, he freely gave them to the world. It was up to later mathematicians to demonstrate them to be correct. "Trust that I'm right" does not fly far in science and academics, nor should it.

Namagiri was a reliable source, but the same cannot be said of insights in general. Jacques Hadamard, a French mathematician who wrote, *An Essay on the Psychology of Invention in the Mathematical Field,* pointed out that most of his subjects went to lengths to verify their insights.[8] 'Trust but verify', supposedly taken from a Russian proverb—*doveryai no proveryai*—is the phrase Ronald Reagan used repeatedly when talking to his counterpart, Russian leader Mikhail Gorbachev, who was bothered by its use since he did not know of the proverb. We are wise to learn from it.

With *doveryai no proveryai* in mind, I return to what my insight told me: mix 5 grams of xylitol in a 45 ml bottle of Ocean. Later, I found out that no one used that much xylitol—the normal amount is less than half that. The difference is important because the lesser amount does not pull any water into the airway, but the amount I used does. That pull is what we know as osmosis—the movement of water through a membrane toward the side with the higher concentration. I had to trust it would work, but I'm a doctor so I was sure to verify. I found out that the movement of water into the airway optimizes nasal defenses—it's *how* xylitol helps our airway defense work optimally While marketed as a nose wash, it's classified as a cosmetic so it cleans, that is not what

8 Jacques Salomon Hadamard. *An Essay on the Psychology of Invention in the Mathematical Field.* 1945.Princeton Univ. Press.

it does. Cleaning removes everything, even the good, beneficial, commensal bacteria living on our nose. What xylitol does is optimize our own airway defenses.

Usually the water in the air we breathe provides enough to optimize defenses, but during winter in the earth's temperate zones, the amount of moisture in the air drops. Cold air cannot hold as much water, so humidity levels drop. Compounding the problem, heating our homes dries out the air we breathe while indoors, making the water available to aid our natural airway defenses drop far below what is needed for our defenses to work. It's why we have a 'cold season'. During summer, we also cool the air with air conditioning, removing water from the air and handicapping our airway defenses in the process.

Nordic countries compensate for this with wet saunas, and in the Islamic world they compensate with ablutions prior to prayers. But places that don't compensate deal with consequences like cold and flu seasons and COVID, a respiratory illness that generally hits the people with weakened airway defenses the hardest.

Five milligrams, the amount in one spray of xylitol, is enough to compensate for the lack of water in the air, and it lasts for six to eight hours in the nose. Using it two to three times per day should be sufficient. And the fiftieth of-a-plum delivered to the stomach is well within the tolerable range. In my drug application for the FDA I told them about its safety. They said they were impressed, but they wouldn't put it in writing.

The voice in my dreams said, "*You have discovered how to optimize the airway's defenses.*"

A Cosmic War

After observing the results with the eye of a physician who had examined many suffering patients, I realized I had discovered how to optimize the airway's natural defense system. Not only did it prevent ear infections, but it also prevented allergic reactions to irritants in the air. And, astoundingly, it prevented asthma, which is often caused by irritants that make it into the nasal passage. A clean nose means no irritants and thus no allergies or asthma.

That was my experience early on, and I wrote up my findings and got them published in a peer reviewed journal, but, unfortunately, in a journal that the national library of medicine does not index, so it's invisible and few know about it.[9]

My efforts to promote a simple study for treating asthma continue to be futile, and while no one admits to it, I am reasonably sure the hesitancy is because the practitioners see preventing asthma as hurting their income, and the researchers see it as destroying theirs.

The guiding voice in my dreams gave me clarity with a simple message: *Western medicine is waging a Cosmic War with microbes. You have found a way to negotiate with them.*

Cosmic wars are not about territory; they are more existential and about deeply held ideas like religion. Reza Aslan's advice in *How to Win a Cosmic War* is key: "don't get in one." The epic importance of the difference between war and negotiation can't be overstated. Pasteur began this war when he described how microbes infect us, erroneously thinking they are weak and defenseless. They are in fact the titans of life on this earth, and

9 The article is available online at Academia and Researchgate. Search "Intranasal xylitol."

our war with them is a cosmic one that we have no chance of winning. They mutate rapidly, and even faster when threatened. They adapt to our weapons, pass those adaptations to their progeny and share them via plasmids with whatever microbe needs them. And they do this with no thought of the intellectual property rights and profit motives that delay our efforts to cope. Whatever we come up with on offense, they can counter on defense and then some. We cannot win this war, and the harder we try, the worse our prospects become.

So that leaves us with Bernard's defenses, Costerton's tools and Ewald's insights as our best options, the aforementioned defense side of the equation that we ignore in our warfare. These pathways implore us to focus on the soil where the microbe is planted and to be kind to our enemies in the process since our enemies can become our friends.

On an individual scale, the microbe's environment is surrounded by our natural defenses. On a larger scale, it's the environment of the host, the host's stress level and how they manage it—stress reduces natural defenses. It's Antonovsky's *salutogenesis* and its foundation in the *sense of coherence*. And on an even larger scale, it's the places where we regain our health: the sanitarium, the Asklepion, and the safe places where we can relax, let go, and play. They all help us stay healthy. They even allow for negotiating with the microbes that either can work for us or against us. The microbes are going to do what they're going to do based on how nature created them. It's up to us to determine the environment where we interact with them and our approach to existing with them.

Evolutionary biologist Paul Ewald tells us how to do it in his book *Evolution of Infectious Disease*. Mutations are random and many of them wind up useless, but there are two pathways to success: virulence and commensalism. The rate of mutation is increased when a threat is perceived. It's the same quorum sensing process discussed earlier. If a microbe can easily get from one person to another it follows the virulence pathway and infects more people and becomes more dangerous to the hosts. If it cannot easily move to a new host, it adapts toward commensalism—to being less virulent. Polluted water promotes virulent cholera. Clean water promotes the more benign *El Tor* strain. No use of condoms allows for the spread of virulent HIV strains. On the other hand, the common use of condoms in Japan is shown to change HIV to a significantly less virulent form. That's commensalism.

Constrain the spread of the microbe and it suddenly has a vested interest in keeping the host alive. The mutations that Ewald is talking about are random, and most are of no help at all, but the ones that follow the paths of evolutionary success are invaluable for what they tell us about blocking transmission. Those mutations are the ones that survive by going through one of the doors Ewald describes: either by spreading with increasing virulence, or by adapting toward commensalism. Ewald shows how blocking transmission puts a very beneficial finger on that scale. Unfortunately, he limits his constraints to those outside the body—masks, gloves, condoms, isolation etc.—consistent with our healthcare system, which likewise ignores our internal defenses while continuing to fight an exclusively offensive war. This is to our great disadvantage because those internal defenses are what Bernard's soil consists of, and optimizing them puts, not a finger, but a foot on the scale—and it's something we all can do.

You cannot win any game, let alone a war, without a defense, and our system ignores most of our defenses like gastroenteritis, the GI tract's washing defense, and rhinitis, the airway's defense, and even mistakes them as illnesses and cripples them with drugs. They are defenses helping us win the war, and they are victims of friendly fire. And there is more to the story.

Let's Play Defense

When a microbe finds its way into our bodies, it must grab hold before it can cause infection. Microbes, both bacteria and viruses, are covered with glycans looking to hold onto the some of the specific glycans that cover the proteins on the surfaces of our cells. Glycans help the immune system identify *us*, the body, as opposed to *them*, the invaders. If we have an organ transplant, we need to turn off this self-identification to prevent the immune system from seeing the '*not us*' glycans on the transplanted organ as an invader and destroying it. For a blood transfusion, we need to be sure to receive our blood type to avoid a nasty transfusion reaction, and it's the glycans on red blood cells that define blood type. For Type O, the glycan is the sugar fucose. For Type B, it's galactose. And for Type A, it's the sugar complex N acetyl galactosamine.

Those different glycans are why we get a transfusion reaction if given the wrong type of blood; our bodies know it's not ours because of the glycans. They are specific, unmodifiable, not patentable, and not profitable for drug companies. They are also immunity's working parts and working with them is a large part of what defense medicine is about.

Many people interested in medicine write about our defenses in terms of our immune system. There are basically two parts to the immune system: innate immunity is what we are born with;

acquired immunity is what we build by overcoming infection. You are born with your innate immunity, and simply put, it's the body's recognition of itself so that it can aim its defensive artillery at the foreign invaders. The lymphatic system together with the glycans are the body's surveillance and identification team, watching for and identifying invaders. The lymphatic system helps us acquire immunity when it recognizes an invader that is not us and the immune system responds.

Glycans are important for another reason as well as reflected in Costerton's earlier reference: they are what microbes hold onto to begin an infection.

If you are at all familiar with medical research you will know how it works with the proteins of both us and our invaders, but those proteins are more properly called glycoproteins, and it's the glycans on the proteins that do the work of identification, which triggers the immune system. But, again, glycans are specific and cannot be modified, patented, controlled and profitable. But the proteins can. Hence, the interest in proteins more than glycans

The other side of the equation is the defensive line the body puts at its openings to make it harder for microbes to gain entrance. Glycans play a large role here as well. Mucins combine our body's signature glycans with protein chains and set up camp everywhere our bodies are open to the outside. Invading microbes can dock on the mucus glycans if we live outdoors, live in a humidified environment, or use and osmotic agent to provide enough airway surface fluid to keep defenses optimal, and if they attach to the mucus, the body says bye-bye by flushing the entire mucin dock, microbes included. But if they dock on the cellular glycans, infection results. This defense is simple, clever, and very effective if not handicapped.

But there is a problem if the microbe grabs hold in a place where the mucus is dry and not optimal, and the cells are exposed—that's when we get sick.

This information is probably new to you—it will be for most readers. It's new because it's not studied, and few people are talking about it. Why the silence? I'll bet it's because there is no money in it. These glycans are all natural, specific in their structure, and neither modifiable, patentable, controllable, nor profitable. The silence is profit-driven medicine at its worst; ignoring simple remedies because they aren't profitable.

But the glycans can be played with and xylitol seems to be the queen on the chess board. Xylitol is in a class of sugar alcohols that have an additional hydrogen atom that makes them flexible. Flexibility allows the sugar alcohol to shapeshift—to look enough like the glycan that it can act like a decoy. The protein binding site on the microbe is specific in its conformation to fit a particular glycan, and an appropriate sugar alcohol can gum up the binding site and prevent it from attaching to its preferred site on our cells. It's a process called *competitive inhibition* and it works! And even though a spray of Xlear has only four hundredths of the amount in a plum in molecular terms there are more than billion billion molecules doing this competing—about $2x10^{19}$ to be more mathematically accurate. That's competitive inhibition in spades.

The availability of sugar alcohols at our grocery stores allows us all to add these commonsense drugs to our defense team and see how easy it is to block microbes from holding onto us, and by doing so we activate Ewald's pressure on them to adapt commensally. This process has been shown to work both in the laboratory with a variety of bacteria and viruses, including the flu virus, SARS, and RSV. And it has been used in clinical studies related to

COVID: in a small clinical study of patients with COVID-19, and a larger one on health care workers in India promoting their own proprietary nasal spray containing xylitol that is not yet available.[10]

Viruses are a bit different however because while bacterial binding sites need to hold on to us, they don't get into our cells. viruses do and they hijack our machinery of glycosylation so they are tagged with the glycans our cells use to identify us—this is the 'remember this' from the earlier quote from *Essentials of Glycobiology*. It's a really important point that no one talks about. It means they look like us to our immune systems, and that's not good. It's why the virus can get in so many places in our bodies and last so long. And it explains why it's so important that our defenses are optimal enough to stop viruses before the get inside our cells.

Is the solution that simple? Yes! It's that simple, but foods that act as drugs cannot make the medical claims that are in fact real. The result is that no one knows what they do.

Whispers of Asclepius

Not everything perceived as coming from Asclepius made sense right away. Some of it may suffer from my own believer's bias. But Ramanujan trusted Namagiri, and I trusted my insight. Again, my wife Jerry was a great help. She became a certified play therapist to

---

10 Damian Balmforth, James A Swales, Laurence Silpa, Alan Dunton, Kay E. Davies, Stephen G. Davies, Archana Kamath, Jayanti Gupta, Sandeep Gupta, M.Abid Masood, Áine McKnight, Doug Rees, Angela J. Russell, Manu Jaggi, Rakesh Uppal. "Evaluating the efficacy and safety of a novel prophylactic nasal spray in the prevention of SARS-CoV-2 infection: A multi-centre, double blind, placebo-controlled, randomised trial." *Journal of Clinical Virology*, **155**(2022): 105248, ISSN 1386-6532, https://doi.org/10.1016/j.jcv.2022.105248.

work with children as they played out the challenges and traumas in their lives under the mantra of "trust the process." Complex problems cannot be analyzed—there are too many variables—you have to trust the process. That mantra permeates what my inner voice told me:

"It is all the same."

It's true that God heals, whether we mean now or when Asclepius is said to have whispered healing thoughts into the minds of the visitors to his temples. And William Thomas's rule from the early part of the last century is just as true now as it's always been: "If men define situations as real, they are real in their consequences." The belief in healing leads to healing. But I found another application for it: the environment and mindset that leads to a healthy individual—which certainly includes but is not limited to their defenses—is the same as the environment that leads to a healthy nation, and everything in between. In other words, it's all the same. We should think of health holistically, accounting for the individual's inner and outer worlds in a coherent way that satisfies and agrees with Antonovsky. It's one where the individual is not afraid of playing with their conditions without the fear engendered by the experts. And a critically important part of this evolutionary play is where one can negotiate with the 'other'—microbes in Ewald's terms, to those thinking differently in the human world—to reach a consensus on the best path forward.

The novelty of these experiences left me with no trail blazed by those who came before me. Like a man with a machete and dense jungle in front of him, I had to hack away one step at a time. And the going was slow. It is hard to practice medicine without realizing that something is very wrong with the philosophies of Western practitioners. And despite shelves full of books and white papers spelling out the problems, nothing changes. In fact, it only gets worse.

Case in point: the National Academy of Medicine (NAM). For decades, their publications have examined and defined the problems, producing thousands of pages of analysis, but as for fixes, all I hear are crickets chirping. This is mostly because our healthcare system is ingrained in our culture and cultures are hard to change. They are especially resistant when a large part of the population profits from the malfunction. Cultures limit our field of view and options for taking action. Cultures do change as new information enters them, but slowly. Radically new information—such as Copernicus' simpler explanation of the observed facts by moving the center of our universe from the earth to the sun; Newton's explanations of gravity and the laws of motion; and the advent of the Internet—forces faster change, but it's still slow by any measure except perhaps the geologic time scale.

NAM hit the nail on the head in their 2001 publication *Crossing the Quality Chasm.* They point to the chasm between the health care Americans pay for and what they receive. We have by far the most expensive healthcare system in the world, yet the measures of our health are mostly consistent with only moderately developed nations. And the chasm has not changed in the 20 years since its publication. Indeed, with the recent declines in American life expectancy, it's getting worse. It's seen as a tactical issue, but it's one of strategy constrained by culture.

Tactics vs. Strategy

Few sentences in my reading of history and current events struck with more force, and led to more contemplation, than the following by Moshe Ya'alon, Israel's former minister of defense. In a comment about Israel's Palestinian policy, he said, "In our tactical decisions, we are operating contrary to our strategic interests."[11]

When it comes to our healthcare system, we, too, are making the same blunder because few of our recommended fixes truly address the problems. The wise Ya'alon was speaking to Israel's strategy with the Palestinians, but the same can be said of our strategy with microbes as well the healthcare system as a whole.

In Israel, military service is compulsory, and most of its citizens are steeped in its warrior culture. When they are hit, they hit back harder—hard enough that the other side will think twice before attacking again. It's a tactical application of the adage, "the best defense is a good offense," and it sums up modern healthcare. We, too, are warriors; our battle is with microbes, the strategy is to hit them hard with antibiotics, and when they resist, we hit them with multiple antibiotics. We bomb the microbes into submission, laying waste to our immune system and spurring a microbial Star Wars program, that, as a Cosmic war we never should have begun.

Pasteur, the man behind pasteurization, placed us on this path by showing us how microbes cause disease. This created a fear of them which brought out the warrior in us. *Bombs away!* This tactical decision was not at all in our strategic interest. Microbes,

---

11 Molly Moore. "Top Israeli Officer Says Tactics are Backfiring." *Washington Post Foreign Service.* (October 31, 2003). Page A01. http://www.washingtonpost.com/wp-dyn/articles/A44374-2003Oct30.html

as previously noted, are far from defenseless. Our war with them is one we have little chance of winning. We need other options, like those discussed here—and we need them in dealing with the system as well.

Defense medicine is a first step in that direction. Nowhere in my medical school training did I learn about Claude Bernard's defensive approach, even though he is honored as the father of modern physiology. To the contrary, the rigors of medical training ingrain Pasteur's fear-based offensive response in physicians. We fight because we are in fight-or-flight mode, responding from an instinctually fearful place instead of a logical analysis of the situation. In doing so, we continue to ignore Bernard's half of what the practice of medicine should be about. We wage war by using antibiotics and antiviral weapons, never looking for ways to help our soil—our bodies and societies—be more resilient. One half of the practice of medicine—about helping our defenses work better in coping with infectious diseases—is never even seen.

This brings us back to why we continue haphazardly bombing when we should be defending as well. The truth is, despite more than a century-and-a-half of knowing better, knowing better is simply not profitable. Knowing better leads to using off-the-shelf, relatively inexpensive natural substances as the first line of defense for the body. It's not to say that synthetic substances aren't effective, but there's an old saying about reinventing the wheel, and it comes to bear in this debate about why we use costly, offense-based synthetic substances instead of the alternatives. We reinvent the wheel because it's profitable!

On the other side, the defensive tools are derived from natural substances that cannot be modified, controlled, patented, and made into FDA-approved drugs. Passing through the gates of FDA drug-hood costs easily over a million dollars, and only then can the manufacturer claim any health-related benefits. The reason the pharmaceutical industry was not interested in Xlear was because if they spent the money to jump through FDA hoops, they would have to charge enough to recoup that money—and that's more than enough to push people into making in their kitchens, leaving the industry holding an empty bag. You know all those advertising claims you hear about what a pharmaceutical drug can do? Not only do you hear no health claims about natural substances, but it's also illegal for the producer to make such claims without jumping through FDA hoops first.

Hippocrates said our food should be our medicine, which is why we call the foods and other natural substances which double as great medicines *Hippocratic drugs.* I'm sure the clever advertisers could get plenty of mileage out of it.

We like simple answers to our problems, and the simplest is "take two and call me in the morning." Pop a pill or two—problem solved. But humans are not simple in our make-up, and neither we nor our institutions can be understood and fixed with simple measures. We, and they, are complex—so is our culture. When people from a western culture go to India to learn meditation and end up getting sick, Ayurveda doesn't work to cure them—they need a pill because their cultural belief is that pills cure. Ayurveda is not real to them in the Thomas sense, but a pill is. That's culture.

It's complicated; it is a matter of knowing all the pieces at play and the connections to track. Working on an algebra problem to find an unknown when there is only one or two to look for is easy. Three unknowns leads to a sharp increase in complexity, known as a "three-body problem." Beer, for example, is a three-body problem since it's made of just hops, yeast, and barley, but when combined they can produce tens of thousands of different beers. In the human body the connections are virtually limitless—it is complex, and instead of simple answers, we get answers called attractors and "basins of interest" and a myopia created by the walls of those basins of interest when we are unable to see over them. Cultural beliefs create walls around us that are hard to see beyond, and in Western culture we have made our problems so complex, the solutions must be nothing less than paradigm-changing. We must completely change how we understand ourselves, each other, and the world.

I found agreement with my ideas among the educators, scientists, and physicians who are members of the International Society for Systems and Complexity Sciences for Health. As an outlier in this group (I'm a family physician with no academic attachments), I presented my ideas at two of their conferences, describing what opened my eyes to Bernard's defensive approach and how it can change the practice of medicine, and how trusting Antonovsky's processes leading to natural health can solve the complexity problem. I outlined Bernard's concept of balance and Antonovsky's coherence and how human beings stay healthy and showed that it is not just a problem of biology, it is family and society, the soil in which the human lives and grows. Beyond biology and medicine we have great contributions from family therapists and sociologists about how to create and maintain healthy homes and communities.

Don't count on medicine alone to come to the rescue—the definition of health is broader and found in the term "well-being," and well-being is mostly about the individual's ability to control the factors that relate to their health where a significant part is financial.

Health is also about balance, and not merely the balance of symptoms in our bodies, but also the impact of our environment on us. It's about financial security because the loss of control seen with poverty cripples well-being. Well-being is much more than prescribing the proper drug and making the proper diagnosis.

This information leads to a new paradigm, a new way of thinking about health. One that applies simple concepts to hard problems. Just as the Copernican heliocentric model explained away the Ptolemaic system, the ability to see over the walls of our basin of interest bypasses much of our complexity.[12] How to do that is the message behind this book and its sequel, and its applications go far beyond individual health and familial resilience. They extend to the living systems that we create: our teams, organizations, and nations. Complex, living organizations share the same rules for fitness as those offered by sociology and family therapy, and their path to well-being is the same as our nation's path, and it's the same as I offer in medicine.

Or as Asclepius whispered, *"It's all the same!"*

The Mind Virus

A new paradigm and a reshaping of our culture are the changes we require. Changing healthcare, particularly in the U.S., will not be easy because it is woven into our culture, yet it can and must be done. My contributions as a physician and inventor were not hard, and I know firsthand how a modern twist on the ancient practice of medicine as taught by Hippocrates enhances patient

12 See "From Microbes to Models" in Joachim P. Sturmberg, Editor. *Putting Systems and Complexity Sciences Into Practice 2018.* (Springer; Cham, Switzerland), and "Salutogenesis Revisited" in J. P. Sturmberg, Editor. *Embracing Complexity in Health* 2019. (Springer; Cham, Switzerland).

health. But I also know all-too well the opposition to change which faces us, and it boils down to one word:

Profit.

The quest for profit is warping the culture of healthcare in Western medicine, and it has been doing so since before most of us were born. The making of healthcare into a commodity rather than a service has compounded the problem. Commodification was a natural result of using insurance as the primary payer of medical bills because insurers needed a standard office call charge—and the office call expanded to become the thousands of medical procedures that are performed in the practice of medicine—and ways to expand their complexity. Billing insurance companies eliminated the patient from the marketplace, removed their function in the marketplace of looking for the best price, and opened the door to escalating profits. This 'show me the money' orientation gives us hypocritical medicine because it focuses on making treatment more complex and profitable. And with insurance picking up the tab the model leads the patient to believe the most expensive treatment is the best, so everyone wants Cadillac care even when the do-it-yourself approach is far better. Simple solutions are ignored, and since they are not profitable there is little research done to confirm them.

Simple solutions cannot be offered through the official channels of the FDA because that path is too expensive, and Hippocratic drugs are excluded. So their medical benefits can't be advertised and consequently few people are made aware of them. Commonsense medicine seeks simple solutions first, primarily through a strategic approach of making our defenses more robust, and educating people about the FDA approved drugs that cripple them.

Shifting healthcare to a better orientation can be done, but it won't happen by convincing the big players to give up their billions. Instead, it will be done through a bottom-up approach—through an educated citizenry who knows better. It is a cultural change that will lead to regulatory changes enabling helpful and accurate information currently excluded.

And one statistic illustrates better than any other why the change is so badly needed. Medical error is the third leading cause of death after heart disease and cancer.[13] The true scope of the problem may be worse when we consider that the mindset of offensive medicine often targets the very defenses the body uses to fight illness. This may be why more people do not recover from heart disease, cancer, and other diseases. It is as if someone wants us to be sick and found a diabolical means of making it happen by working from within the healthcare system. It is not an enemy doing this. It happens because of our focus on offense. This focus has burrowed so deep into how we think, it cannot be differentiated from the thought processes which are intrinsic to the practice of medicine. We have been infected with a fearsome virus of the mind which has altered the very way we think about health and medicine.

13 Martin Makary, "Medical Error—the Third Leading Cause of Death in the U.S.," *British Medical Journal* 353 (May 3, 2016): 2139. https://doi.org/10.1136/bmj.i2139.

CHAPTER TWO:

Surviving the Next Pandemic

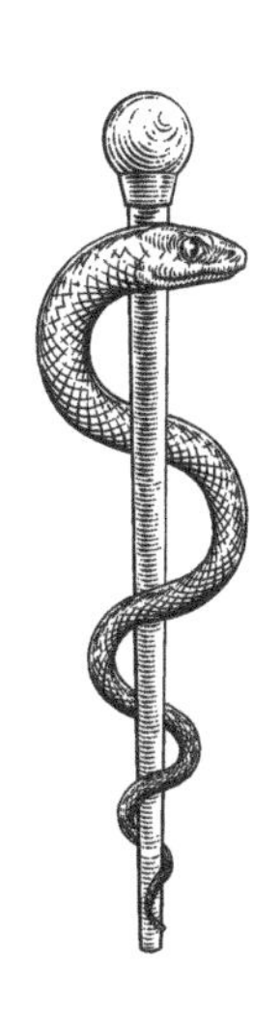

Our offensive use of antibiotics gave me an immediate connection with what Ya'alon means by tactics. Beginning with Pasteur, our strategy was to kill all the microbes with antibiotics, an offensive practice that includes the overuse of antimicrobial cleaners and soaps. That's a big part of our culture due to advertising, but that strategy is shifting as we see the result in more resistance and learn the value of microbes for protecting us from infection, aiding our metabolism, and guiding our development. Half of our body weight is microbial—it's the result of, and the continuation of, the processes of endosymbiosis that began it all (more about that later). Many of these helpful microbes are in our gut, and they are hit first and hardest when we use oral antibiotics, so they are a major source of antibiotic resistance. Isn't it obvious that our overuse of antibiotics is a tactical decision that's contrary to our strategic interest?

Awareness of the major downsides to this strategy is spreading, and oh do we really need a cultural change and a paradigm shift to support it! As Ya'alon implies by his statement, proper strategy needs a proper culture. Culture is the nervous system of the body politic. Its health is our health.

And it's how we survive the next pandemic. We coexist with our microbial neighbors. We make them our friends and learn from their example of how to organize as a community, a nation, and a people.

Most epidemics today happen from airway contamination. In the past it was more GI, from unpasteurized and contaminated milk or water with microbes like cholera. There was also in the southern parts of the U.S., an early outbreak of malaria. That outbreak, as well as the yellow fever problem in those building the Panama Canal were also solved by public health measures, by addressing the mosquitos in Panama and using screens in the south.

The common element in all pandemics is environmental exposure to pathogens. The interesting point about this is that our social pandemics also begin the environmental exposure to pathogens, seen in society as things that cause outrage or fear in individuals. Social pandemics spread from person to person, just like infections, and they spread far faster today due to social media. Try to imagine how big the 1938 War of the Worlds panic would be if they had Twitter. On second thoughts that may not be a good example; if they had Twitter far fewer would listen to the radio and hear the broadcast. QAnon may be a better example for today.

But as interesting as the commonalities are, the question is dealing with the next infectious pandemic. First off it is more likely to be an airborne infection that remains asymptomatic for several days so that tracing contacts is impossible. Even COVID accomplished that. So, when you inhale such a microbe the first line of defense is where it goes and what it meets on the way. A graphic portrayal of what it meets is available on You Tube that you can find by searching "mucociliary," "humidity, " and throw in "Fisher and Paykel"—the New Zealand firm that made it available. Of course, they have an interest in putting it out; they make humidifiers for CPAP machines. There are numerous videos showing how the mucociliary process works, but this one is the only one I have seen that shows what happens when the humidity drops—within ten minutes the flow of the mucus stops.

Humidity measures the water content of the air outdoors that we have adapted to over past millennia. Until recently, like 5,000 years ago, we lived outdoors where the humidity rarely drops below 40% unless you live in a desert. The video shows that the optimal humidity in the nose is 100%. This level is maintained because the moisture in the air we exhale is already there. The needed extra must come from the air we breathe in. What this moisture feeds is the airway surface fluid, a layer of water on top of our cell surfaces that provides the necessary elements for the cleaning to work. It's deep enough that the cilia can sweep in it, and it has enough water so that the dry and concentrated mucus made by the goblet cells can absorb enough to become the viscous, sticky and slimy mucus that works. Without adequate humidity the whole process stops.

So, if you live in an artificial environment with central air and cooling, like most people in our country, your airway defenses are likely compromised. What your invader meets is dry mucus with plentiful spaces where it can get to the target glycans it's looking for on your airway cells or cilia. You get sick. In the case of viruses like SARS-CoV-2, their preferred target is the glycans on the ACE-2 receptor that hangs out on the cilia. It's an easy target if the cilia are not sweeping. From their docking places they can easily migrate down to the cell, gain entry and multiply, and hijack our own process of adding glycans to their surface, so they look like us. Our immune system no longer clearly sees them as invaders. They are like spies. This is why this virus can get into so many places in our bodies, cause such a variety of symptoms, and last for so long.

If, however, you work outside, or you wear a mask that traps your exhaled moisture and feeds it back when you breathe in—resulting often in 100% relative humidity[14]—or you have enough

14 Joseph M. Courtney, Ad Bax. "Hydrating the respiratory tract: An alternative explanation why masks lower severity of COVID-19." *Biophysical Letter,* March 16, 2021; 120,(6):994-1000.

indoor plants to keep your humidity in the optimal 40 to 60% relative humidity range, then your mucus is flowing. Interestingly a recent study of Toronto's homeless population, done to show the increased incidence and need for care, looked at 736 people with an incidence of 47% who had COVID infection—higher than the national average of ~35%—but none died or required hospitalization.[15] Of course the authors did not explain those benefits; in fact they didn't even mention them. You don't need to use an osmotic sugar alcohol spray if you can do without it by being homeless and living most of your day outdoors where the humidity is optimal, but if you can't, do what you can to optimize this defense. When the defense is optimal the mucus will trap and remove most of the pollutants that enter our noses. It does this with microbial invaders mostly because the mucus is filled with the same glycans that the microbes are looking to dock with. In the case of COVID it's our ACE-2 receptors.[16] The glycans in our mucus are easy to detach so when the virus docks on one it is cut loose and washed out. On the way out, it is run by the lymphoid cells and other elements in the airway like the adenoids where the process of acquired immunity is begun. If your nose is optimal, you gain immunity without getting sick, and that's how it should work for everyone. Whatever microbe can infect us must be able to dock on one or more of the glycans on our cells. *If they can't*

15 Richard L, Nisenbaum R, Brown M, Liu M, Pedersen C, Jenkinson JIR, Mishra S, Baral S, Colwill K, Gingras AC, McGeer A, Hwang SW. Incidence of SARS-CoV-2 Infection Among People Experiencing Homelessness in Toronto, Canada. JAMA Netw Open. 2023 Mar 1;6(3):e232774. doi: 10.1001/jamanetworkopen.2023.2774. PMID: 36912833; PMCID: PMC10011938.

16 Chatterjee M, van Putten JPM, Strijbis K. Defensive Properties of Mucin Glycoproteins during Respiratory Infections-Relevance for SARS-CoV-2. mBio. 2020 Nov 12;11(6):e02374-20. doi: 10.1128/mBio.02374-20. PMID: 33184103; PMCID: PMC7663010.

dock, they can't infect us. If you are one of those who needs the sugar alcohol to make this work, *and you are using the right one, you get an added benefit, described by Costerton, in that the sugar alcohol is likely blocking the binding site on the microbe that it uses to dock on its preferred glycan.* When this happens the microbe can't hold on anymore and is attenuated, and it is just as safe as all other attenuated microbes in our vaccines that we use to build immunity. This process is how we can spur microbes to be more friendly to us; it's what Ewald says happens when we make it harder for the microbe to get around.

Keeping your nasal defenses optimal is a simple way to reduce this problem as well as cope with all airborne epidemics—and sugar alcohols help by mimicking the glycans so that the microbes dock with them instead of us. In this way it is possible and realistic to see this as negotiating a way out of a cosmic War we cannot win.

Another way they could help, not considered by Costerton, is if you load the environment with them so that they latch on to the microbial protein and disturb structure and function. An example of this is what happens to a diabetic's hemoglobin molecules when their glucose gets too high. We have all seen the TV ads for drugs that lower A-1-C. The A-1-C is the measurement of how much glucose is on their hemoglobin. Most glycosylation is done enzymatically inside the cell, but when glycans are numerous they can do the same thing on their own. Glucose is also one of the glycans and when it is abundant it clings to the hemoglobin molecule, which makes it harder to release the oxygen it is carrying to our cells—its function is disturbed. The same kind of thing happens when microbial structures are covered with other numerous glycans or the flexible sugar alcohols that can mimic them. Uhari and his group in Finland fed the microbe behind most ear infections xylitol and saw the bacterial cell swell and get contorted. They explained this by saying the microbe got indigestion because it couldn't digest the xylitol, but it could be that xylitol attaches to the glycans on the microbes surface and distorts the cell and its function just as glucose does to hemoglobin.

These concepts of defense medicine and negotiating with microbes are both new and groundbreaking. I wish I was able to point to an "ah ha" moment when it all became clear to me, but that didn't happen. It came in bits and pieces, and the periodic insights were also a source of wonder. It really is the beginning of a paradigm change that revolutionizes how and what we see.

CHAPTER THREE:

Where this goes

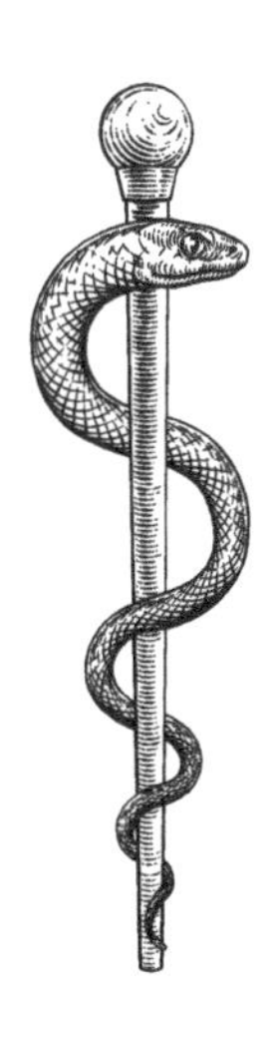

My Master's degree in the history of science and ideas played a profound role in my medical education and practice in part because the study of science examines how the current paradigm—our model of how the world works—affects our thinking. In history, for example, Leopold von Ranke was a nineteenth century historian who thought he could write history "wie es eigentlich gewesen war"—as it actually was. He thought and wrote in the Newtonian paradigm of mechanics, assuming a clear connection exists between action and reaction. It works for explaining the mechanics of baseball and planetary orbits and how birds fly, but history is not mechanics, and his idea was a pipe dream because history is never as simple as action and reaction.

It is true for medicine as well. I studied osteopathic medicine because my wife and I wanted to raise our family in a small town, and osteopathic medical schools train more family physicians. Andrew Taylor Still founded the osteopathic practice of medicine in the 1870s as a revolt against the humoral practice (think: bloodletting). His reforms dealt with the structure and function of the body and was entirely consistent with Newtonian mechanics. He used skeletal manipulation to correct its structure, enabling better functioning. As with history, however, the human body is much more complex than structure and function, and action and reaction.

Osteopathic medicine enjoyed some notoriety during the flu epidemic of 1918 because the mortality rate of patients treated by osteopathic physicians was 1/40th of those treated by mainstream physicians.[17] That benefit was attributed to skeletal manipulation

17 R. Kendrick Smith, "One Hundred Thousand Cases of Influenza with a Death Rate of One-Fortieth of That Officially Reported under Conventional Medical Treatment," Journal of the American Osteopathic Association, 1920; 19:172-75.

and the value of addressing structure. Eighty years later, Dr. Harold Magoun showed that the benefit was more likely due to osteopathic physicians not using drugs to reduce fever.[18] He recognized fever as one of our natural defenses. It is useful in fighting any infection.

Magoun tried to move the paradigm to the one that I call defense medicine. It recognizes that we are more than machines in need of tactical balancing. Sometimes we need the imbalance of a fever—it honors the complexities of the human condition.

The process of healing, like the fever, is often disruptive, and it's easy to see how the disruption could be considered an illness. But our defenses do not need *treatment*, they need *support.* They are evolutionary-derived and treating them with drugs cripples them and eliminates the survival benefit behind why we have them in the first place. That's why more people died from the flu when given fever-reducing drugs.

Animal studies bear out the finding: artificially-infected animals treated with fever-controlling medicines die more often than those that aren't treated. This way of thinking opens the door to a new paradigm of medicine that goes beyond just changing a practice or two. And it is not just easy to recognize and understand; it's also an essential first step in changing the paradigm.

COVID created a pandemic like the 1918 flu, and there are ways that we can help our own defenses cope with this virus that are more strategic than tactical so that we can survive the next pandemic. Implementing them may even result in surpassing the forty-fold survival benefit seen by earlier osteopathic physicians.

18 Harold Magoun Jr. "More About the Use of OMT during Influenza Epidemics," *Journal of the American Osteopathic Association,* October 2004; 104, No.10: 406-7.

Offense-based tactics that focus on killing them aren't doing the job against microbes, and there's no way to win the cosmic war we created against them. We need different thinking that is beyond the Newtonian model.

Even with computer modeling, complex living systems can't be analyzed with Newtonian mechanics. With apologies to Jacques Monod, the Nobel Prize winner for his work in cellular biology, who notably said, "the cell is a machine, the animal is a machine, man is a machine," there are way too many interconnections and networks in our bodies that defy that simplicity.

And it's why our 'drugs' are such a problem. Pharmacologists isolate and analyze a section of our chemical physiology to find a new drug but ignore many interconnections in the process. As a result a promising new drug may come out and seem like a miracle, then a few years down the road we find out that the adverse side effects are worse than the benefits. How many more people have to suffer and die before we learn that good tactics don't make for good strategy, and that ignoring our complexity always has unintended consequences? One way around this problem, that would also be far less expensive for the pharmaceutical industry, would be to measure drug safety by looking at the best measure of health: life expectancy. Giving a new drug to rats for two or three generations would show any problems, and if there were any questions doing the same thing on rabbits should resolve them. If a drug passes this test it should be safe enough for our Public Health Service to use and determine efficacy—if we had a functioning PHS.

And a little child shall lead them

I'm amazed by how children can simplify complexity. While not exactly a little child—she had three of her own at the time—one of my granddaughters did that for this book when she summed up its complexity with her statement:

"Grandpa, it's like the warp and the woof."

The what and the what?

The warp are the threads that run lengthwise through a fabric, and the woof (more commonly called the weft, but we both like dogs) are the threads that run across it, like the lines of longitude and latitude on a map. With fabric, the warp and the woof need to be equally strong to prevent the fabric from fraying. Apply the idea to understanding the fabric of society and it gives us a wonderful analogy.

What she was looking at with her warp and woof was Darwin and evolution. The warp is the natural selection that blesses us with beneficial mutations that spread gradually to extended progeny because they allow for greater reproductive success. The woof is what Darwin couldn't pin down. He too labored under the Newtonian paradigm and with time it was easy to see the action and reaction of natural selection; he could explain it in *On the Origin of the Species.* But it was different for the woof, even though he spent much of the rest of his life working on it. The woof is why civilization exists. Why do we cooperate as we do. He wrote a lot about sympathy, in much the same vein as Adam Smith did with his *Theory on Moral Sentiments* a century earller. But he could not work his ideas into Newtonian clarity so he didn't describe why we have civility. But the woof is real—it's just complex.

The warp and the woof are survival strategies in the Darwinian sense, and ideally they complement each other. But instead Darwin's warp of natural selection has been hijacked by Social Darwinism, the mistaken idea that natural selection is the reason for the great and powerful, of survival of the fittest that pleased the hierarchy of the day and gave faulty reasons for their power and authority. The resulting warp of survivalism and self-interest is our dominant paradigm, and it is ripping apart the fabric of society.

Seeing how my granddaughter beautifully simplified a complex sociological problem got me thinking about how children adapt to the complexities around them. They use trial-and-error methods to find what works. We call it play—take things apart, put them back together, build towers, wipe them out, build families with dolls and animals—and it's how children learn; it's the scientific method adapted to childhood. Open-minded inquiry imparts a Newtonian paradigm that everything functions mechanically. For most of us, that way of viewing the world continues to dominate our thinking long after childhood. We take things apart to fix them. We look for and find connections, and they become real in their consequences. The 3 Rule strikes again!

The Rule may be why placebos work. Belief becomes reality. In the case of a placebo, the belief is built on the idea that pharmaceutical drugs cure medical problems. Researchers identify the mechanism for how a drug works, and it does its job as expected. The belief works powerfully in the minds of patients. But it's not a mechanical process. The mechanical paradigm is no longer sufficient, because Newton's paradigm cannot explain living systems. If we continue to use it to fix our problems, we will do more harm than good.

Darwin exemplifies this shortcoming. The causative and connective steps of evolution are the natural selection of beneficial mutations that aid survival. They are easily understood in the mechanical Newtonian paradigm. But when Darwin tried to apply the same reasoning to explain the rise of civilization, which he wrote about after *On the Origin of Species*, he could not describe the reasons under the Newtonian paradigm. The social elements that lead to civilization, such as sympathy and human relationships, are not mechanical, and they evade simple cause-and-effect connections. Their complexity demanded a new paradigm, and he did not have one.

I agree wholeheartedly with the biologist Theodosius Dobzhansky's conclusion that nothing in biology makes sense except in the light of evolution. And yet, that bright light has never been able to penetrate the complexities of human cultures and civilizations. But those complexities can be understood if we look at the archetype of a complex adaptive system: a little child. No other organism adapts into such a wide variety of geographic and ethnic environments. Children reveal a simplicity beyond the complexity, and this understanding opens a doorway to addressing all aspects of well-being.

This sort of paradigm shift enables us to see over the borders of our basin of interest, and, like children, see things for what they are and call them out for what they are not. Offense medicine is flawed and not the only game in town. Fact is that much better options are staring us in the face. We will plumb them for answers as we continue to explore real solutions that will fix our healthcare system. It is the path I followed, and it is simple to see once the doorway opens.

A Paradigm of True Greatness

In the American historian of science Thomas Kuhn's classic book about paradigms, *The Structure of Scientific Revolutions*, he tells us that a paradigm is the mental model we build in our brains to explain the world, and that a paradigm shift is a shift from an old model to a new one. But those models are basins of interest, and those stuck in them often have a bunker mentality. Their basins become foxholes that protect them, and are difficult to see beyond, especially when working inside one makes your work more profitable. It is in your own self-interest to vigorously defend your position, and your paradigm.

In healthcare, it is our situation today. We all see that the system is deeply flawed and in trouble—except the people dug in too far down to see, the ones benefiting from the system and its disfunction. At the same time, too few people are looking for answers outside their paradigmatic boxes, and their proposed solutions are, more often than not, tactical ones that don't strategically address the causes.

Harvard professor Clayton Christensen in *The Innovator's Prescription* sees healthcare needing *innovative disruption* to control the costs, and to repair and improve it. His solutions, however (such as the use of more surgical centers) are tactical and do not reach beyond his basin of interest as a business expert. His solutions are adjustments, not innovative disruptions. Solutions need to be as great as the problems they are meant to solve and the questions they are meant to answer—they need to be strategic, not tactical, and, since the problem is cultural, they need to be innovative at that level. At the same time, however, innovative disruptions need to build better systems—they are not destructive.

And if they are they need to be, in Joseph Schumpeter's terms, creatively destructive.

But to our great misfortune we are stuck culturally in the model of Social Darwinism, the false application of natural selection to social systems—so the rich and powerful should be in control. In that model, social structures, like governments, are seen as constraining self expression, they should be allowed to die. That was the platform of Donald Trump and his advisor Steve Bannon: drain the swamp; get rid of the regulatory state; cut off the life support and see what happens. The thinking behind that model is all-too common, but few exemplified it and promoted it more than Newt Gingrich, who gleefully acknowledges that he paved the way for Trump. As a child, Newt wanted to be a zookeeper and preferred to be called "Newt" in reference to the salamander, rather than "Newton", his given name. He read Darwin and mistakenly grabbed onto Social Darwinism so he thought that politicians should adopt the survival of the fittest model.

Trump ran to 'make America great again', but his idea of greatness relates to his wealth and power—the survival aspects of Social Darwinism. In his mind, and in the minds of many people like Newt who view the world similarly, the acquisition of wealth and power is a sort of triumph of natural selection. It is the view of the alpha whose prowess entitles them to a greater share of the spoils. But the reason that natural selection is so successful is that it shares the profitable mutations. And if you take the natural bacterial nature of this sharing, it's done freely and wildly with no thought of return on investment. There is no need for a Gini coefficient (a statistical measure of inequality) in nature; all members of the ecosystem are equal. And the fittest ecosystem, again, is where diversity is greatest. and the interdependence of the members is optimal. Wealth and power define survival in our profit-centered culture, but they have nothing to do with what Darwin means by survival of the fittest—a term, in fact, which he did not use.

Fitness for Darwin was a species or system affair. It was never about the individual, or who could crush the competition on their way to hogging the wealth and power. Add to this his thoughts about the evolutionary advantages of caring relationships and cooperation and you have, in fact, quite the opposite. The fittest are those who can strategically share resources amongst their group, because as a group they have better odds for survival.

Trump's greatness in the Social Darwinist sense (which, again, is not Darwin's idea) is a powerful attractor in a complex society. It is about the ego and the desire for wealth and power. But egos have a downside. When they get bruised, they react from their midbrains, their lizard brains, and they react tactically, without strategic thought for the larger picture. This is a very powerful response. It began with the earliest animals with brains, the reptiles. It provided a rapid response to threats that saved many lives—and we have all inherited it. And it applies to our egos just as much as it does to our existential threats. And if there's nothing to stop men or women with powerful egos, they can swallow entire nations. But this thinking has a downside because the midbrain responses are limited to those of the lizard: fight, flight, or freeze.

Here's something to think about regarding people with big egos: They are called narcissists after Narcissus of the Greek legend. Before he was born, his mother visited a prophet, Tiresias, about the child she was carrying, and Tiresias predicted that her son would thrive as long as he did *not know* himself. 'Knowing oneself' was prized in ancient Greece. It was the first of the three maxims in Apollo's Temple, along with 'nothing to excess' and 'certainty brings ruin'. A narcissist sees only the warp aspects of their personality. A person who knows who they are can see both their warp and woof. They can see themselves as an independent individual with character and assets, or they can see themselves as a part of their community. It's the second aspect that Tiresias honored and if a narcissist sees it they may stop being a narcissist.

The more I thought about it, the more I realized that individuals in nature are not that important—it's more important to contribute to and be a part of your society, your ecosystem. It resonates with Robin Wall Kimmerer's idea that we are "human delegates to the democracy of the species."[19]

That's a long way from Social Darwinism, a concept which came from sociobiologist Herbert Spencer's egocentric application of evolution to our social framework. He argued that family, wealth, and position—major determiners of one's place in society—amount to the beneficial mutations behind natural selection. That kind of survival of the fittest is another word for hoarding and has nothing to do with evolution or nature.

The *survival* side of evolution is better seen in random mutations that spread slowly through the sharing of genes, and their survival advantage means we live longer and reproduce more, so they are shared more often, and eventually we all have them. Random mutations continue today, and it takes time before we find out if they are helpful or not. We need to know sooner rather than later if we are to face the next pandemic with the best defense we can field.

19 Robin Wall Kimmerer. *The Democracy of the Species.* (New York: Basic Books, 2021)

What Survival, and Surviving a Pandemic Really Means

Until recently we were stuck with having to live with the harmful mutations behind the microbes that cause our pandemics, and finding ways of treating them. CRISPR technology, though, allows us to play with our genes to correct the mutations that are not helpful—but only if we are sure that the gene is isolated and has not been integrated into other aspects of our body's complexity. Anyone who understands complexity cringes at the term "junk DNA" because it demonstrates appalling ignorance. We have given the godlike power to alter the fundamentals of creation into the hands of people who think like Monod—that man is a machine that can and be safely tinkered with when problems are found. These apparent simple solutions to complexity miss the point of complexity: you can't mess with one element without disturbing the whole—it's the argument of the vaccine deniers, and I tend to agree with them, in large part because natural immunity, as described above, is so easy, much more effective, and far safer. And this holds true for social pandemics as well as infectious ones.

All too often our responses to both medical and social problems is to treat the symptoms; it's tactical, not strategic. The mRNA vaccines for COVID are a tactical attempt to replicate a part of our natural immune defense, but by introducing a foreign protein into our cells the process is asking for problems; it's adding something unknown to a complex system. And just like the viruses that hijack our glycosylating processes in order to look like us, these novel and foreign stalk proteins are similar glycosylated by our cells and can escape much of our immune system. Those benefiting reap the profits and the government covers the unintended consequences. In much the same way financial wizards play with the regulations and wind up with crises, as in 2008, or bank failures, a frequent and recurring problem in our history, and the government picks up the tab. The intentions are generally good, it's just that they are short on strategy. The Supreme Court decision in Brown v. Board of Education concluded correctly that 'separate but equal is inherently unequal', and the decision to integrate the schools was unanimous and well intended. But a few people, like Thomas Sowell, saw the forthcoming problems in that tactical decision: it went against the culture and the culture responded by creating private, often church related *de facto* segregated schools. The tactical response looked at separate schools. Another approach, more consistent with the culture, would be to address the unequal part of the equation by ensuring adequate funding for all schools rather than relying on property taxes, which guaranteed their inequality.

Our expanding understanding of our world is like random mutations. They can grow into novel innovations, like nuclear power, or disruptive ones, like atomic bombs. Like random mutations they are both a blessing and a curse, but there is no putting Pandora back in the box. Random mutations that work have two avenues, as in evolution, and as described by Ewald: they are either more virulent—warpish and surviving by inflicting hurt on others—or as commensals—woofish and cooperating with their hosts. As we learn more we can make our lives better by pushing virulent organisms—including humans—toward the latter option, and the same strategy of social isolation works well with warpish people.

This understanding is a powerful asset for increasing our evolutionary fitness, which includes that of our civilizations. Sharing innovation in the spirit of cooperation and empathy is behind Darwin's rise of civilization. While being opposite in nature, these two evolutionary innovations—survival of the individual by novel mutations, and the cooperative communication behind group fitness—make up Indra's net, a Hindu symbol for the fundamental unity of everything. They are what we call healthy pathways, and like the warp and the woof, both need to be equally strong to hold together the fabric of civilization.

In our self-centered and competitive model of society many disparage the value of cooperating. Evolutionary theorist Lynn Margulis studied early bacterial life on earth and shows its value. She proposed a theory about the symbiotic relationship between two types of bacteria and larger cells, and the theory is now largely accepted. More than a billion years ago, a single bacterium had a beneficial mutation that enabled the use of sunlight to free oxygen and make sugars. Down the line, it or an offshoot joined with a larger cell to become the intracellular chloroplast that powers all plant cells. Another bacterium that learned how to break down those sugars to make energy joined with a larger cell to become the intracellular mitochondria that power all animal cells.[20] In this cooperative leap, bacteria with these differing abilities combined in a process called endosymbiosis. The resulting enhanced cells are the parents of our world of plants—powered by chloroplasts that produce the oxygen we need—and animals—powered by the mitochondria that turn the sugars we eat into energy.

Survival—the warp—helps the individual in small steps. Cooperation—the woof—leads to progress in leaps and bounds.

Endosymbiosis continues today as bacteria find commensal niches in our bodies that aid our development and growth—and defense medicine spurs that process.

---

20 Lynn Sagan, "On the Origin of Mitosing Cells," *Journal of Theoretical Biology* 14 (1967): 225–74.

Like Indra's net, the two pathways—survival and cooperation—weave together to make up the fabric of our lives and our world. Both the warp and the woof are strong.

Our social fabric, on the other hand, is stretched terribly thin due to either our inability or outright refusal to see beyond the paradigms shaping how we see the world. That of Social Darwinism has eclipsed with self-interest the wide and free sharing that is fundamental to the survival aspects of the natural world. We are in survival mode and unable to evolve to cooperation mode. Our political impasse, where one side seems unable to see any value in the other, is a great example of this myopia. Like different paradigms, the two sides of U.S. politics are incommensurable. The word incommensurable is related to commensal, which is what we term the mutually beneficial relationship between friendly bacteria and ourselves. For a long time we did not see any microbes as friendly or beneficial. They too were incommensurable, but that view is changing.

Here's to hoping we can change our political views as well, perhaps through a process of evolution as we learn and internalize the value of cooperation and its evolutionary benefits. Both views—mutations that benefit the individual for the warp, and community, empathy, and cooperative communication for the woof—are a part of nature, and both are necessary. There should be no wall between them. They should be understood, and subject to compromise.

The *greatness* we really need is not that of size and power, but the evolutionary fitness that comes from the *cooperative communication* that enabled our ancestral *homo sapiens* to win out over the Neanderthals. The path to that greatness is the effort behind consensus and cooperation and the ability to work together that led to the success of *homo sapiens* in building communities and civilizations. It's built on the meaning, understanding and control that Antonovsky saw in healthy individuals. It's not found in ego trips that lead to a society of self-interested individualists who, on their way to the top, will step on or over everyone and anyone in their way.

In environments where the fittest survive, great size and power are more often signs of less fitness. It's like the quarter horses, inbred to run a quarter-mile race, that give out after three-eighths. And the Mongolian horses, bred to go long distances, that run a 25-mile race and die from a heart attack a quarter mile from the finish (something I witnessed happen). And the dinosaurs that grew so great in size that their environment, clobbered by an asteroid, could no longer support them. These are cautionary tales. In evolution, the fit survive, not the great, and fitness comes with both helpful mutations from the individual, improving it from the diversity, and cooperative sharing with the community—and ideas are the same as mutation. The fittest are actually the most cooperative because cooperation enhances survival. Innovations are shared. But Darwin's theory got warped to support the view of a capitalist paradigm where "share" is a dirty word.

It began with Spencer's Social Darwinism and continued in Ayn Rand's objectivist philosophy and the libertarian principle behind self-interest in our marketplace. If we can learn anything from evolution, however, it is the assertion that when self-interest dominates, it leads to a community population with the least fitness. For a society, the individual is the source of the mutations/ideas (the warp), the diversity cross-pollinates and expands their usefulness, and the cooperation (the woof) is what makes the fitness. These two opposing elements must be in balance in order to make a strong fabric. This concept extends to our society and our politics as much as it does to biology and weaving fabric.

Take the United States for example.

With the exception of Thomas Jefferson, who felt that the blood of patriots was needed periodically to replenish the passion for freedom, most of our founders feared the prospect of repeated rebellions, and the democracy behind them. The Shay's Rebellion in Massachusetts where a group of farmers protested their debts—caused in some part by Massachusetts not accepting the Federal script that was their pay for the Revolutionary War—took place just a year before they gathered in 1787 for the Constitutional Convention in Philadelphia. Right from the start their fear of democracy weakened the woof of the social fabric they were creating by limiting the right to vote to property holders like themselves. That move shifted the alignment of the U.S. toward the warp of self-interest, and the imbalance between the warp and the woof created weakness in society's fabric. Wealthy plutocrats took advantage of the weakness, and it has only been getting worse as individualism, competition, and profits outweigh community, cooperation, and charity.

It did not have to end up this way. A very similar rebellion occurred in ancient Greece, but the governing leader, Solon, forgave the farmer's debts—a move that those studying the period see as giving birth to Athenian democracy. While our founders feared democracy they did understand the principles and included them in them in the Constitution's Preamble. They are predominantly citizen-oriented:

> "form a more perfect union."
> "establish justice."
> "ensure domestic tranquility."
> "provide for the common defense."
> "promote the general welfare."
> "ensure the blessings of liberty to us and our posterity."
>
> In theory, a government founded on those principles should create the balance necessary for a healthy nation. The equal strength of the warp of self-interest and the woof of collective interest creates a strong fabric for society. But in practice, self-interest won out, and the result is the fraying of social institutions.

Our health is following the same general decline for the same reasons. We are out of balance. There is too much emphasis on the profitable warp of offense medicine at the expense of defense medicine, and as a result we are unraveling. Our well-being is suffering. It happens when self-interest overwhelms collective interest, and our society has so strongly shifted to self-interest that we are now further than ever from Aristotle's goal of creating a government which serves the *collective* interest. In the coming pages we will speak more about how to rebalance, revealing specific ways we can fix healthcare and ourselves. The body politic is not the only one suffering from this imbalance—our physical bodies suffer as well.

Changing the culture is central to this task, and defense medicine opened my eyes to the solution. What began as a physician's quest to do his job better took me on a path to seeing far beyond the traditional definitions of medicine and healthcare. Defense medicine's applications of moving control down, negotiating more, getting out of lizard brain tactical responses, and striving for consensus are universal for all complex and adaptive life, including that of our nation, and it is an easy concept to see once the window opens. Just ask my granddaughter, who grasped it quickly and intuitively. Despite the fierce and well-funded resistance of the old paradigm, this simple logic will catch on quickly.

It really is all the same.

The Defensive Front Line

Medical training is all offense. You learn about various diseases and the microbes that cause them. Then you learn about the "weapons" at your disposal—the various drugs, treatments, and procedures—and you "go to war" against the "invaders" and "enemies." By the time you come out of medical school, you are a gladiator in the arena of healthcare, and your job is to attack, attack, attack. Fortunately, my training in osteopathic medicine and its emphasis on structure and function left the door open to framing the practice of medicine in terms of defense, but even before my medical school training I knew this war should be more than just an offensive one. Thank goodness for my familiarity with Claude Bernard's idea to strengthen our body's natural defenses and resilience, especially at its openings where we are most vulnerable and defenses are most robust—and, unfortunately, when primary defenses are overwhelmed the secondary can be most bothersome as well, which aids their misinterpretation as illnesses. He rightly deserves the title of "Father of Defense Medicine."

While oral rehydration started me down the path of defense medicine I got a lot of help. Biologist George Williams and psychiatrist Randy Nesse looked at some other bodily defenses including fever in *Who Gets Sick: The New Science of Darwinian Medicine.*[21] It may have provided the insight, discussed earlier, for Harold Magoun, DO, to take another look at the use of fever-reducing drugs by osteopathic physicians during the 1918 flu pandemic. A fever is an inflammatory response that helps healing and it's the first response of our body to injury or infection. But in medicine we are taught to fear such responses because they may cause seizures. Febile seizures are more common in children, and I remember my pediatrics professor discussing their treatment with the shouted words: "For heaven's sake, don't just do something, STAND THERE!" In short there are few long-term effects of a febrile seizure. If unsure of the cause, or while treating a child less than two years old, a visit to the doctor to search for a site of infection is a good idea because some of them—rare but serious enough to think about—can be in the brain, and lethal. There are a number of things we look for in the emergency room to rule out a serious problem when a patient is feverish. But the fever is much more likely to be beneficial than harmful.

Williams and Nesse relate a variety of animal studies done with artificially infected animals, both cold and warm-blooded. If they were prevented from developing a fever, either by blocking cold-blooded animals from warming in the sun, or by using drugs to reduce the fever in warm-blooded ones, more of them died. Doing such experiments on humans is unethical. But we don't seem able to learn from animals by incorporating these fundamental insights into the practice of Western medicine. Magoun's article applied this to humans, but the connection with our defenses and

---

21 George C. Williams and Randolph M. Nesse, *Why We Get Sick: The New Science of Darwinian Medicine* (New York: Random House, 1995).

defense medicine was not spelled out. Even today few are aware of the benefits of a fever. A fever of 39°C (102.2° F.) "boosts the protective heat-shock and immune response (humoral, cell-mediated, and nutritional) whereas ≥40°C (104° F.) initiates/enhances the antiviral responses and restricts high-temperature adapted pathogens, e.g., severe acute respiratory syndrome coronavirus 2 (SARS-CoV-2), strains of influenza, and measles."[22] Likely the best treatment for a fever is oral rehydration to keep everything functioning well and provide for the perspiration that cools us off.

The inflammatory response is a major defenses. For an injury like a sprain or fracture, the inflammatory response fills the injured space with blood, providing support, or immobilization, and causing pain that sends the message to give that part of the body a rest so it can heal. Bolstering the defenses through the use of splints and crutches allows us to tune down the inflammation naturally with ice and elevation. If the problem is an infection, the inflammatory response brings blood and all its cellular immune substances for fighting infection to where the infection is. We should learn that lesson for all our defenses—bolster them; don't shut them down.

When you shut down defenses, you increase mortality. That's the nature of defenses; we have them because of their survival benefits. Cripple those defenses and you lose the benefits. As harmful as it is to cripple the inflammatory response, for example, Western medicine did that for close to 3,000 years with the practice of bloodletting. For all that time the dominant model in medicine was the humoral system where imbalances in the

22 Singh S, Kishore D, Singh RK. Potential for Further Mismanagement of Fever During COVID-19 Pandemic: Possible Causes and Impacts. Front Med (Lausanne). 2022 Mar 2;9:751929. doi: 10.3389/fmed.2022.751929. PMID: 35308547; PMCID: PMC8924660.

humors (blood, phlegm, bile, and black bile) were the causes of disease, and the treatment was aimed at rebalancing them. And the symbol of medicine, what is today the stethoscope, was the knife and bleeding basin. In this system the signs of "too much blood" (in quotes to signify the error of this belief) were those of the inflammatory response: redness, pain, fever, and swelling; or *rubor, dolor, calor, and tumor* in the humoral lingo. And the treatment, of course, was getting rid of some of the blood. And it works to reverse the symptoms; you can literally watch your patient *look* better as the loss of blood leads to the shock response where peripheral blood is shunted to the central organs needed to sustain life. The redness turns pallid, the pain is lessened, they become cool and clammy, and if there is any swelling it too diminishes.

With that visual confirmation it is no wonder that bloodletting as a medical practice lasted as long as it did. You could literally watch as the signs of too much blood disappeared. Then and now, balancing symptoms is how we measure health, and our thinking is still humoral. The illusion of improving health through visual confirmation of the reduction of symptoms is likely why it took 3,000 years to stop bloodletting, and today it's the same with a lab test. Just as culture trumps strategy, shock trumps inflammation as it shunts blood away from the periphery to the central organs needed to maintain life, and what appears to be good medicine is actually killing the patient.

In the culture of medicine, bloodletting is one of many practices that cripple the body's defenses, and they all end in the same result. More people die because we eliminate the survival benefits associated with the defense. The idea is expressed in the gallows humor of a poem that parodies Dr. John Lettsome, the Revolutionary War-era physician:

I, John Lettsome,
Blisters, bleeds and sweats 'em.
If, after that, they please to die,
I, John Lettsome.

John "lets 'em" die, and we assume it wasn't on purpose—physicians of the time grew accustomed to their patients dying, which is of course part of medical practice, but you could say that patients were lucky to survive after the good doctor got his hands on them and subjected them to his treatments! The practice of medicine has come a long way since then, but we continue the same sort of rationale with the tradition of shutting off bothersome symptoms that also happen to be defenses. We got rid of the humoral model and replaced it with scientific medicine, but we didn't change the thinking. We got rid of the humors, but we still try to balance the scientific humors we replaced them with. We did this in 1918 by treating the fever. We continued it with COVID by treating a fever greater than 101°. Sometimes the body needs to be out of balance—like when it's being invaded by microbes. There is an emerging model in healthcare that recognizes this—it's called the allostatic model—it's based on our responses to stresses so it accepts out of normal responses. It slow going and would work better if it expanded to include an understanding of our natural defenses.

Most of our infections enter the body through one of its openings—the nose, mouth, birth canal, and excretory openings. These vulnerable openings are also where our defenses are most robust, and when the primary defenses are insufficient, the bothersome symptoms of their backups let us know that something is wrong. But our wrongheaded healthcare system views the bothersome symptoms as illnesses.

Bloodletting was an outgrowth of *humoral medicine.* While we say that we no longer practice that brand of medicine—we're more scientific—the mentality of the culture lives on. We say we are scientific, but the most fundamental concept in the practice of modern medicine—balancing the body—is primitive, and we are long past the time to shift our attention to the ways we can honor and bolster the body's defenses rather than cripple them. That is the defensive half of the equation we are missing. And it is a game-changer.

As discussed earlier, biologist Paul Ewald proved that microbes become more deadly when they can easily infect a new host. If, however, easy transmission is blocked, the microbe must live in its current host, so it adapts toward commensalism.[23] We use hygiene methods like clean water, bed nets, hand-washing, and condoms for blocking transmission. Diseases are shown to be less severe when these methods are widely used, but unfortunately the methods stop at our skin, and we do not consider the natural defenses in the soil—the body and the environment it lives in—where the microbe is planted.

Take tooth decay, for example. It is the most common infectious disease suffered by mankind and an excellent example how epidemics begin and how we can help our defenses prevent it. It's also the science behind xylitol's original medical uses. Until the 1970s, we did not see tooth decay as an infectious problem, but we knew it was associated with dietary sugar. Association is not causation, but Finnish dental researcher were suspicious enough to look. They conducted a two-year study by dividing participants into three groups and giving regular table sugar to

23 Paul W. Ewald. "The Evolution of Virulence." *Scientific American,* April 1993. Pp. 86-93.

one group, fructose to another group, and xylitol to the third group. Then they tracked what happened to the teeth of people in each group. As you can imagine, the sugar eaters showed the greatest progression of tooth decay. Fructose eaters showed less decay, and the xylitol eater showed no progression—the acid making microbes were tamed or gone and decay reversed in some cases.[24] Turns out, xylitol binds with calcium ions, leading to remineralization of tooth enamel.

That was the initial thought, but it went further—much further. The study spawned a new industry making gum sweetened with xylitol—gum being a good way to regularly deliver xylitol to the teeth. An industry that tried, like I did, to get the FDA to enable them to make the beneficial claims behind this food. They too gave up, for the same reason—too expensive without a way to keep the cost within reason.

And naturally, while they knew what xylitol did, they wanted to know how it did it. They found the cause of tooth decay: bacteria living in tooth plaque—the biofilm on our teeth—that eat dietary sugars and produce an acid that attacks tooth enamel. In time, a cavity forms. But only certain bacteria are the culprits behind this infectious process of tooth decay. The main culprit, *Streptococcus mutans,* they found is affected by xylitol, and it was immensely gratifying when a pioneer behind the Finnish dental studies, Kauko Mäkinen, agreed with my ideas about its mimetic and competitive action with the glycans.

Many follow-up studies confirmed xylitol's benefit in preventing decay, but you won't find "xylitol saves teeth!" splashed

24 Scheinin A, Mäkinen KK, Ylitalo K. Turku sugar studies. V. Final report on the effect of sucrose, fructose and xylitol diets on the caries incidence in man.

across the packaging of gum sweetened with it. Why? It's a drug claim that needs FDA approval, and unfortunately, xylitol is a natural product available at many grocery stores, offering no way to control and patent its use and make it profitable enough to jump through FDA hoops. Manufacturers cannot advertise the factual claims. Few people know about xylitol's dental benefits. Dr. Catherine Hayes of Harvard School of Dental Medicine calls this block to public health education unethical.[25]

A study conducted in Belize gave us much more information about how it works and its value. It was done in two parts; the first compared gum sweetened with xylitol or sorbitol for young children and found xylitol to be a much better choice.[26] The second, a follow-up study four years later, sought to find out what happened to teeth during the intervening time with no xylitol consumption. Significantly, researchers found a relative absence of cavities in the permanent teeth that had erupted during the earlier study. These teeth had 90% less decay than the neighboring teeth *in the same mouth.*[27] The only possible explanations of this finding are that either the bacteria causing the decay are gone or they are transformed. Under the influence of xylitol, the bacteria actually do both! It also shows the stability of biofilm and that it can be very localized and not affected by even a close environment.

---

25 Catherine Hayes. "The effect of non-cariogenic sweeteners on the prevention of dental caries: a review of the evidence." Journal of Dental Education, October, 2001; 65(10):1106-1109.

26 K. K. Mäkinen, C. A. Bennett, P. P. Hujoel, P. J. Isokangas, K. P. Isotupa, H. R. Pape Jr., and P. L. Mäkinen, "Xylitol Chewing Gums and Caries Rates: A 40-Month Cohort Study," *Journal of Dental Research* 74, no. 12 (December 1995): 1904–13.

27 P. P. Hujoel, K. K. Mäkinen, C. A. Bennett, K. P. Isotupa, P. J. Isokangas, P. Allen, and P. L. Mäkinen, "The Optimum Time to Initiate Habitual Xylitol Gum-Chewing for Obtaining Long-Term Caries Prevention," *Journal of Dental Research* 78, no. 3 (March 1999): 797–803.

It is a demonstration of how xylitol and other sugar alcohols can negotiate with microbes. The study also shows us that biofilms are stable—at least they were for the four years between the studies. That means that if you have this kind of plaque on your teeth, you want to keep it. Your bacteria are protecting your teeth. Keep them safe. Feed them more xylitol to keep the message strong. Think twice about going to the dentist for teeth cleaning that tries to remove the plaque, or at least educate the hygienist that you have good plaque and to go easy on it. Scraping it all away may actually be harmful in the long run.

That story carries over to the other places where we have biofilms. Wounds create biofilms as they heal, but these biofilms are filled with pathogens. Randy Wolcott is a wound care doctor who began using xylitol on his patient's wounds and found they healed faster and better. Word got around quickly, too. At the arena where a couple of my grandkids were showing their horses there was a horse in a sling being treated for a barbed wire interaction. When I asked what they were using on the wound, the answer was xylitol.

In all these interactions the simplest explanation is that xylitol has an action on how the microbe attaches, and that mechanism explains how the microbes causing tooth decay are neutralized. It's consistent with Ewald's explanation for how commensalism ensues after microbe transmission is made difficult. Interfering with adherence applies pressure there as well.

The bottom line is that xylitol protects your teeth. To get the best protection, use it often, in granular form to sweeten your coffee and tea, and as an additive to gum, mints, toothpaste, and a variety of candies. And if you have kids make sure they get exposed early, when their teeth are erupting, both primary and secondary, in order to get the long lasting benefit of healthy plaque. A drawback of xylitol being outside of FDA control is anyone can label their product as containing it, but in what amount? To be useful orally, a product needs at least a gram of it per serving. If it's listed early on the ingredients list, you can at least be assured that the product contains a greater concentration. My son's company created Spry gum, mints, and toothpaste to meet this need.

Nathan Sharon also wrote about bacteria and teeth; he wrote that lectins have a sweet tooth.[28] Lectins also play a role in glycan binding. They are proteins that bind with specific sugars and most of them are the glycans, so lectins play a very powerful role in studying glycans. He was also a pioneer in the study of glycans and how microbes attach to them. The glycan he studied, mannose, is not particularly sweet, but it is a sugar, and it is common among the glycans that help the body identify itself.[29] He studied its interaction with *E. coli*, the bacterial culprit in most urinary tract infections (UTIs). They attach to the mannose that is prevalent on the surfaces of the genital tract. The home for these bacteria is in the GI tract, and in cases of UTI, the road to infection starts there. His discovery inspired him to experiment by feeding the patients D-mannose to treat UTI, and the results were essentially the same as what xylitol does for teeth: bye-bye infection. The GI bacteria that causes UTI choose mannose as their dance partners, and in their embrace they dance their way right of the dance hall: the body excretes and replaces them with bacteria that don't hang onto the mannose and therefore do not cause infection.

---

28 Sharon N, Lis H. Lectins--proteins with a sweet tooth: functions in cell recognition. Essays in Biochemistry. 1995 ;30:59-75. PMID: 8822149.

29 N. Sharon, "Bacterial Lectins, Cell-Cell Recognition, and Infectious Disease," *FEBS Letters* 217, no. 2 (June 15, 1987): 145–57.

Sharon spent his life trying to get this point across, but it fell on deaf ears because surely, good medicine could not be so simple. But follow-up studies proved Sharon's discovery, and today it is common knowledge among alternative doctors that treating UTIs with D-mannose is just as effective as antibiotics and with fewer side effects. It is even more effective than antibiotics at preventing the recurrence of infection.

Another side to this story came with Sheryl King, a veterinarian at Southern Illinois University. She was looking at the problem of uterine infections that sometimes follow artificial insemination and wondered if Sharon's findings would help. She tried it in the lab and found it did; infections were blocked when the mix included mannose. But she was surprised when she tried it on horses—none of them were infected, but none of them were pregnant.[30] When she looked at it under the microscope the sperm were seen surrounding the bacteria. The acrosome of the human sperm, the part that attaches to the egg, is essentially blocked by mannose. Mannose in glycan chains are often where the chains divide into two so there are two binding areas on the mannose molecule, one for the microbe and one for the sperm. There are now several patents approved for the use of mannose as a contraceptive. I think they would be safer than any of the FDA approved drugs that make up our common contraceptives, but the patents are not of great value either because the expense of drug hood is prohibitive—just like Xlear and the dental uses of xylitol. If you want to try mannose as a contraceptive buy some

30 King SS, Speiser SA, Jones KL, Apgar GA, Wessels SE. Equine spermatozoal motility and fertility associated with the incorporation of d-(+)-mannose into semen extender. Theriogenology. 2006 Apr 1;65(6):1171-9. doi: 10.1016/j.theriogenology.2005.08.002. Epub 2005 Sep 8. PMID: 16154188.

mannose—available on line—and some capsules. Fill a capsule with the mannose and insert it early in foreplay.

Commensal Biofilms to Invaders: 'No Vacancy'

The benefits of defense medicine in the war on microbes are clear by now. Returning to the coronavirus pandemic, a recent study shows that xylitol, sorbitol, or erythritol—simple, off-the-shelf sugar alcohols commonly used as sweeteners—prevent the microbe from taking up residence in the nasal and pharyngeal (throat) tissue. With all of these being effective to varying degrees Costerton's three methods all play as possibilities, but we do not yet know exactly how it works. If Occam's razor still holds and the simplest explanation is the most likely to be correct, then Costerton's third, interfering with adherence, is the answer. As with the Belize dental study, it tells the microbe to "shape up or ship out." And every one of the billion-billions of xylitol molecules doing the negotiating repeats that message. A healthy biofilm can say "no vacancy here" to strange microbes, but xylitol can get into biofilms, like it does in wounds, and make them even healthier. While the glycans on our cell surfaces are fixed in their form so that our immune system and the microbes seeking a docking site can recognize them, the sugar alcohols are flexible. They can shapeshift to fit into the microbial "hands" seeking a glycan as a place to dock (like a key into a lock), and while they may not be *the key* that precisely fits into the microbial lock, they are close enough to gum up the works. It's still competitive inhibition, but with a decided advantage—the shapeshifting sugar alcohol adds to the glycans on our mucus or cell surfaces in providing billions of docking options other than infecting us. The microbe attaches to the decoy, and you do not get sick.

The commonsense drugs to fight viral infection are available at grocery stores. Adding a teaspoon of an appropriate sugar alcohol to a small bottle of saline nasal spray; using it two or three times a day should keep your nose clean and free from SARS-Cov-2. But there is scant clinical evidence for this claim, and I don't think it will take long for you to figure out why. This simple, off-the-shelf solution for preventing microbial infection can help us all, but it doesn't reward the system. And our healthcare system is designed to seek only the solutions and treatments that are the most profitable.

The preventative benefits of sugar alcohols provide us with worthwhile ideas on how to negotiate an end to our offensive war against microbes. Thinking in terms of Bernard's soil and defense medicine, they are the defensive line, but they have teammates. Biofilm is another important natural defense.

As a child playing with a microscope, John William Costerton, the same Bill Costerton discussed earlier, wondered why he saw no bacteria in stream water, but just below the water he observed a smooth film covering the rocks in the stream bed, like what Anton Van Leeuwenhoek, the father of microbiology and developer of the first microscope, observed about dental plaque. Both saw that the film is composed of countless microorganisms—bacteria. Biofilms are important in medicine for several reasons. Most importantly, they protect us. Many say it begins before birth, but whenever it begins, our bodies are doing a symbiotic waltz with countless tiny bacteria. They are a part of us, and like endosymbiosis, they play a role in our development, even in our emotional growth. Biofilms in our GI tracts and elsewhere protect us from many pathogens just by being there and putting out their "No Vacancy" signs. They also help by producing many of the messenger molecules that our body uses to coordinate itself.

Unfortunately, some medical approaches take the "nuke 'em" approach and end up wiping out this protection; for example, douching as treatment for vaginal infections. Bad idea because it wipes out the primary defense: the lactobacilli in the vaginal biofilm. Those helpful bacteria make the environment mildly acidic and thus hostile to most pathogens, and they produce antibacterial substances that prevent infection from common pathogens. Douching washes out these helpful bacteria and eliminates their natural protection.

Douching is no longer a preferred treatment for vaginal infection, but the paradigm remains. We continue medical practices that essentially douche the respiratory tract—as with nets pots and other large volume washes. As stated earlier Xlear® is marketed as a nose wash, but that is not what it does—it optimizes our own mucocilliary cleaning, it tweaks rather than changes the airway environment. Commensal bacterial biofilms at all of our openings are a major part of the body's passive defenses and a significant part of all primary defenses. All of them work 24/7 without producing symptoms. If we can optimize all our primary defenses we are home free—and it is possible.

When primary defenses are not up to the challenge, secondary ones come forward like linebackers and safeties on a football team, and they produce the symptoms we experience as bothersome. These symptoms tell us something is wrong that needs help. The nausea, vomiting, and diarrhea symptoms associated with the GI tract, and the runny nose, coughing, and sneezing symptoms associated with respiratory tract challenges are wrongly labeled "gastroenteritis" and "allergic rhinitis" and are bombed into submission with a variety of drugs and treatments. But these symptoms are the body's natural responses to pollution and 'the solution to pollution is dilution'. The body washes itself, but its washing does not disturb the commensal biofilms: leukorrhea is the washing of the vagina; gastroenteritis is the washing of the GI tract; and rhinorrhea is the washing of the airway. The symptoms are bothersome but necessary. Suppressing them only makes matters worse.

As a first line of defense, the body's resident biofilm of friendly bacteria hangs a "No Vacancy" sign and backing them are the glycans on our mucins providing a docking space and thereby preventing invaders from establishing a beachhead on our cells. Acids are another backup for preventing infection. The GU (urinary) tract is mildly acidic, provided by the hydrogen peroxide made by the lactobacilli that also favors their growth, and the GI (intestinal) tract is highly acidic, to break down foods into the smaller molecules we can absorb and use. The acid and the friendly commensal bacteria are its primary defenses.

The primary defenses in the respiratory tract are the combination of the local biofilm, the mucus that holds onto the pollutants, and the microscopic hairs—the cilia—that sweep it all out, down the throat to be recycled in the stomach. They are a defensive team, and when optimized they work very well together.

Talal Nsouli is the ENT who took care of President Clinton's allergies, and he wanted to know if defenses similar to those of the GU tract are inside the nose. If so, would the use of a Neti pot—an "aggressive lavage tool," like douching—wash out the biofilm and the mucins, crippling that part of the defense and increasing sinus infections? And the answer was yes.[31] Inside

31 Dr. Nsouli presented his report at the Annual Conference of the American College of Allergy, Asthma, and Immunology in Miami Beach, Florida, on November 9, 2009. The report is available online at https://www.medpagetoday.com/meetingcoverage/acaai/16870. Accessed January 1, 2019.

the nose, the friendly bacteria and their biofilm, along with the mucins and the cilia are major defenses, not just for the nose but the entire respiratory tract—and the airway surface fluid is critical in this cleaning process, and enough moisture in the air is crucial in keeping it optimal.

In an ideal world, this process works perfectly, but we do not live in an ideal world. We have 'cold season' when colds and respiratory infections are common because the defenses are not working. And it should tell us that we should be doing something differently since this happens every year when we turn the heat on and reduce the moisture needed to keep it working.

It is indeed no coincidence because our noses have adapted over time to pull the moisture needed for the airway surface fluid from the humidity in the air we breathe. When we heat that air we are reducing the amount of water that our noses can pull from it. Relative humidity is our measure of air moisture. Outdoors, it usually measures in the range of 40-60 percent, which has been shown to be the Optimum Zone, while indoors in our comfortable heated homes and workplaces it commonly drops to 20 percent or less. At that low level, you get shocked when you touch something metal after shuffling across the carpet, and your mucociliary clearance defense dries up and stops working.

We know 40 to 60% relative humidity to be optimal because Anthony Arundel and his colleagues looked at the health effects of indoor humidity and found it.[32] Spending enough time outdoors or with fresh air indoors gives nasal defenses what they need. It is likely to be the major reason we are seeing a protective benefit

32 Arundel AV, Sterling EM, Biggin JH, Sterling TD. "Indirect health effects of relative humidity in indoor environments." *Environ Health Perspect.* 1986 Mar;65:351-61. doi: 10.1289/ehp.8665351. PMID: 3709462; PMCID: PMC1474709.

from the SARS CoV-19 virus by being outside. This virus starts as a respiratory infection, and as you can see from the oval on the chart these infections are rare when the humidity is over 50%. Our bodies adapted to living outdoors in optimal humidity. Optimal humidity lessens all respiratory problems, both infectious and allergenic. That is what our ancestors adapted to. They did not have epidemics until cities were built, where lots of people in a relatively small living area contributed to their spread. But along with cities came homes—enclosed spaces—and fireplaces that warm indoor air and lower the relative humidity, crippling our defenses and allowing for epidemics. The opposite of those infections and epidemics is what the chart indicates when it says, "there is insufficient data for respiratory infections when the humidity is above 50%".

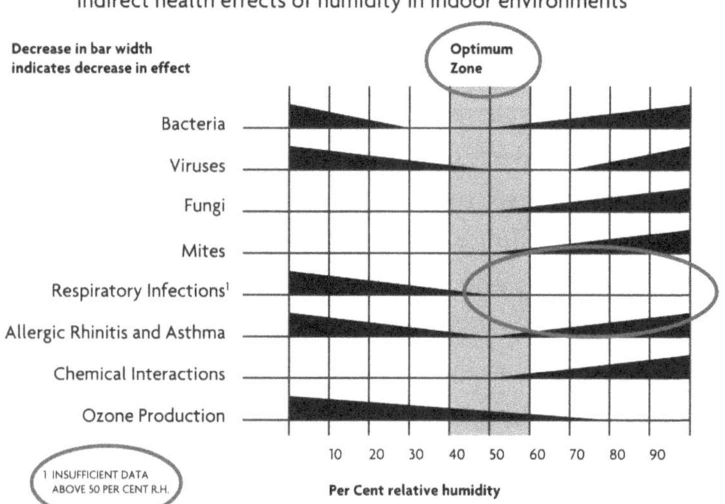

Optimum relative humidity range for minimizing adverse health effects.

Are Simple Defenses too Simple?

If our healthcare system were more interested in preventing illness than profiting from it, we would hear a lot more about simple preventions like humidifying air to counter the dehumidifying effect of heating it, and working with glycans that cannot be patented and profitable. During the coronavirus pandemic, public health experts would tell people to go outside and get some fresh air, but no one seemed to connect that advice to how it works—it optimizes our airway's defenses. Nor do they advise quarantine and isolation with a humidifier. Nor do they humidify hospitals where critical patients are treated with respiratory therapy rather than optimizing their own defenses. The evidence for the efficacy of these simple treatments is so obvious, it makes one wonder, why aren't they in widespread use? Is it too expensive? The data show without a doubt that optimal humidity aids our primary respiratory defenses. And if the primary defense is optimal, we have no need to break out the big guns when symptoms appear. Our primary defenses work 24/7, symptom-free if we are fortunate, and when needed we get the backup defenses like a runny nose or fever, which tell us that our primary defenses are stressed and need help.

But no, we go ballistic over a runny nose and use a palette of drugs to turn off the faucet, and in a sense it is no better than bloodletting. We are the collateral damage in the misapplication of offense medicine. More people die, and it is staggering to think that increasing survival does not trump every other consideration. You mean, the practice of medicine is no longer about maximizing survival? I can hear Hippocrates asking, "Then what sort of medicine do so-called modern people practice?"

We should be looking instead to ways of aiding our defenses. If all games have two sides this is one side that is ignored in our health care.

Oral rehydration therapy (ORT) is another example of playing good defense. Nausea, vomiting, and diarrhea are not bothersome symptoms to treat, but our immune system's attempt to wash out the problem.

In the GI tract, pathogens trigger the symptoms, and at their most severe, as in cholera, they lead to death by dehydration. But keeping the body's fluid tank full through use of ORT prevents dehydration and virtually eliminates mortality. The body's natural defenses can wash out the bugs and their toxins, and the person recovers, usually taking just a day or so longer than when antibiotics are used. You could say that defense medicine involves keeping your players fresh and supplied with everything they need to do their job as best they can for as long as necessary. What more can you ask of a defense?

In life, just as in sports and games, there is offense and defense—and in order to win, you need both to perform their best. Cholera easily bypasses the GI defense of stomach acid, and the resident biofilm is not much help, either. Cholera kills us, not by infecting us but by our immune system erroneously thinking that the microbe's toxin is much more of a threat than it actually is. This mal-perception triggers such an enormous response from our immune system's washing defenses that we die of dehydration. ORT resolves that problem simply by keeping your tank full. The proper balance of salt and glucose activates the sodium-glucose transport system in the upper intestine. One molecule of glucose and two of sodium together work to pump 264 molecules of water into the body.[33] The ORT solutions promoted by the World Health Organization around the world approximate these proportions.

Josh Ruxin, who wrote a history of ORT, thinks it is ignored in the U.S. because of the profit-oriented interests of healthcare. He argues that utilizing it would save close to $1 billion dollars a year in the U.S. alone.[34] He may be right, but people in one basin of interest often cannot see over the walls of their paradigm. Viewing ORT as nothing more than an interesting treatment for cholera allows for ignoring it because our IVs can do the job just as well and are much more profitable. That was also Ruxin's conclusion. as to why it's not used more.

An element he did not include is the simple fact that ORT is a mixture of ingredients in our cupboard. Anyone can easily mix it up in the kitchen, so it lacks the financial benefit that pharmaceutical companies desire as well as the ability to make the drug claims that are clearly appropriate. Patentability and the stamp of approval from the FDA allow the manufacturer to tell people what their drug does. It is the same reason we can't advertise the dental benefits of oral xylitol, nor those following optimizing our respiratory defenses. This regulatory obstacle is a major problem in our system because it leads to ignorance about powerful and effective commonsense Hippocratic drugs.

> The recipe used in Bangladesh to treat cholera is easiest: a pinch of salt and a fistful of sugar in a glass of water, and drink more than you lose.

---

33 Meinild A, Klaerke DA, Loo DD, et al. "The human Na+-glucose co-transporter is a molecular water pump." *Journal of Physiology.* 1998 Apr 1; 508 (Part 1):15-21.

34 Joshua N. Ruxin, "Magic Bullet: The History of Oral Rehydration Therapy," *Medical History* 38 (1994): 33–97.

The evidence we have reviewed is easy to find, and it put me on the road to arguing for a new paradigm in medicine: defense medicine. And it's so simple, even a child can understand it. Summed up, defense medicine means, as much as possible, to promote and optimize the body's defenses with inexpensive Hippocratic drugs rather than crippling them with FDA-approved drugs.

I once talked to a researcher looking at deaths from toxigenic *E. coli* and asked him if anyone was looking at the use of drugs that stop diarrhea—the natural defense—in those that died. Doing so would allow the bacteria to remain in the body longer and find more ways to penetrate. He told me that no one was interested in that question, i.e. there's no money in it. And the same old reason is to blame: the question opens the door to using our natural defenses, but our war with microbes is only offensive. Asking it leads to a paradigm shift that is consistent with Ruxin's argument about why IVs are preferred by mainstream medicine over ORT: it is not in the interest of the dealers who manage our profit-oriented system.

This brought me around to our warfare with microbes and brings us back to a point I made earlier that bears repeating. The war began when Louis Pasteur blamed what were presumed to be simple, weak, defenseless, microscopic bacteria as the cause for our diseases. That view was wrong on all counts, but it made a declaration of war seem reasonable. In reality their numbers, variety, and ability to adapt make them formidable. For an estimated two billion years early in our world's history they were the only living organisms. They learned how to recycle, and that is their main task today and even our bodies are seen in that light. The war against microbes is unwinnable. They mutate and develop resistance far faster than we can keep up, and they share their mutations, both vertically by dividing and horizontally in plasmids (small packages of DNA or RNA), freely sharing their mutations with other microbes. And they do it with no concern at all for profits or intellectual property rights.

Our microbial partners on this earth are more like the mythical Titans and our war is a cosmic one, and the only way to win a war is to avoid it in the first place. Too late for that, but the door is now open to negotiate a truce. That is what defense medicine is all about, and Western physicians should take note.

Again, there are lessons here from history. I knew of the importance Bernard placed on the quality of the 'soil' in which a microbe is planted, but Pasteur turned the microbe into something to fear. Our fight-or-flight response led to the war with microbes, and our warfare-focus has blinded us to the more peaceful options that are easily found in defense medicine.

I see parallels between this thinking and our current focus on treating symptoms. When I mentioned this to a colleague, he replied that we are still practicing humoral medicine because our thinking has not changed—it is part of our culture. And without a change in our thinking we wind up making similar mistakes. Outdated strategies drive our tactical decisions, and our culture constrains our strategies, preventing us from changing the paradigm.

Changing the paradigm will happen when we realize that the game we are playing with microbes has both offense and defense—and playing defense is just as honorable as offense, just as effective, far less expensive, and not harmful to our protective biome that helps us to stay healthy and develop optimally. That is clearly true with using ORT to support GI defenses. It is just as true with using xylitol to support our respiratory defenses. But it is a cultural change when one sees it as *negotiating* with our microbes. It is a paradigm change, just as the voice in my dreams told me.

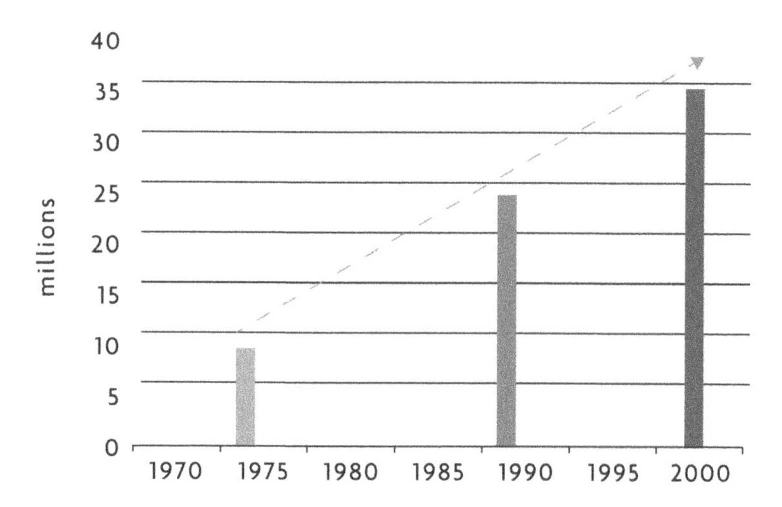

CURING ASTHMA

Supported by the success of ORT and the nasal use of xylitol in my medical practice, I began looking deeper into our defenses and finding ways of helping them. Upper respiratory problems, such as childhood ear infections, are the most common complaints seen by family physicians. For the past half a century, they have been increasing at an unmistakably steady rate, increasing two and a half times from 1975 to 1990, an increase of 5% annually (see chart to right).[35]

Asthma, too, is rapidly increasing. Defined by a constriction of the lower airway, asthma is not considered to be an upper respiratory problem, but upper airway irritants often trigger the bronchospasms that cause lower airway restriction. Asthma

35 Schappert SM. "Office visits for otitis media: United States, 1975-1990." *Advance Data*, 1992 Sep 8;(214):1-19. PMID: 10126841

is a disease of the airway and there is only one airway. Autopsy findings in people known to have drowned showed that in many cases their lungs were dry. The conclusion of the pathologists in this study was that the water in the upper airway triggered the bronchospasms that prevented them from inhaling the water.[36] They may have drowned but the cause of death was an acute asthma attack. The bronchospasm of asthma is part of our airway defense system that recognizes something dangerous in the upper airway and prevents it from getting to and damaging our more vulnerable lungs. Bronchospasm tries to prevent that from happening. Histamine is the trigger for this backup defense, and bronchospasm is listed as its fourth effect, along with increasing the water supply, making more mucus, and causing sneezing as part of the cleaning. All those benefits and defenses are what antihistamines stop.

Beginning in the early 1970s, something started causing asthma rates to rise. Researchers tracked asthma-related hospital discharges in Charleston, SC, showing level numbers in the decades leading up to the 1970s and a steady rise afterward, with a steep rise among the Black population. Lacking any other model than blaming the patient the authors concluded that it was related to obesity.[37] But the paradigm of defense medicine clearly shows different causes behind the dramatic rise in asthma rates demonstrated by various studies.

One might look at the delivery of noxious gases to the black

36 A. R. Copeland, "An Assessment of Lung Weights in Drowning Cases: The Metro Dade County Experience from 1978 to 1982," *American Journal of Forensic Medical Pathology* 6, no. 4 (December 1985): 301–4.

37 D. D. Crater, S. Heise, M. Perzanowski, R. Herbert, C. G. Morse, T. C. Hulsey, and T. Platts-Mills, "Asthma Hospitalization Trends in Charleston, South Carolina, 1956 to 1997: Twenty-Fold Increase among Black Children during a 30-Year Period," *Pediatrics* 108, no. 6 (2001): E97.

communities that were first disrupted by the Interstate Highways. Or the fact that by the mid-1960s, half of all new homes came equipped with central heating and air conditioning, and that percentage continued to rise during the subsequent decades. Central air necessitates those homes be better sealed from outside air, and consequently the combination of airtight sealing plus artificial heating and cooling decreases the indoor humidity to less than the optimal 40 percent, crippling respiratory defenses. But wait, it gets worse with the widespread use of cold pills—the combination of antihistamines and decongestants—first developed in the 1940's to stop runny noses. They became available over-the-counter in 1965. Decongestant drugs shut down the taps in the lining of the nasal membranes that are needed to be open to replace the water in the airway surface fluid, and histamine is the trigger for the back up washing. These pills effectively cripple our airway's defenses, just as bloodletting did to the inflammatory response, and marketers didn't waste a minute before saturating the public with exhortations to use them. Plus, Medicaid provided them as samples, explaining to some extent the spike seen in severe asthma cases among Charleston's Black population.

No one is interested in looking at life expectancy in the users of cold pills, but the early information on their use in the annual *Physician's Desk Reference* showed they were accompanied by more respiratory infections. The FDA came close to this in 2007 when they saw some deaths in infants and children associated with cold-pills and persuaded the manufacturers to remove their pediatric formulations from the shelves. They thought the problem was due to overdosing. When I called them and suggested it was because these pills crippled an evolutionary defense they said, "that's an interesting idea."

Saline nasal sprays became popular soon after this. They helped sooth some of the resulting dryness but did not seem to prevent infections. But adding a sugar alcohol like xylitol did the trick. I soon discovered that, in the right proportions, the benefits went beyond the simple miracle of curing my granddaughter's ear infections. A great-niece with asthma was able to play basketball and do gymnastics without her asthma medication, and a grandson, allergic to his grandmother's cat, was able to visit without his usual facial swelling. Sinus infections disappeared just like the ear infections. In essence, all upper respiratory problems were significantly reduced, if not eliminated. I had found a way to augment, and optimize, our respiratory defenses.[38]

Xylitol is a Hippocratic drug. It is "food as medicine" in the Ayurvedic tradition. Medicine is anything that helps us deal with a medical condition. Hippocrates and Hindus view food as medicine, but in the U.S. and elsewhere we have shifted away from simple solutions and toward a morass of government regulation that inserts unnecessary complexity into health care and makes the pharmaceutical industry incredibly rich and powerful. But then again, the alternative at the turn of the 20th century was worse.

How We Got Here

Early in our history, "patent medicine"—proprietary medicine made and marketed under patent and available without a prescription—hit in waves like plagues. The free availability of arsenic, cocaine, heroin, alcohol of dubious origin and other dangerous

38 A.H. Jones. "Intranasal Xylitol, recurrent otitis media and asthma: Three case studies." *Clinical Practice of Alternative Medicine*, June, 2001;**2**(2):112-117.

compounds raised serious public health concerns but made people feel better and put money into the pockets of the sellers. Their abuse was legal and mostly tolerated, and their public availability was promoted by the warp culture of self-interest. In the late 19th century, Oliver Wendell Holmes Sr., one of the most beloved physicians of his time and the father of a future Supreme Court justice, said: "I firmly believe that if the whole materia medica [all substances used as remedies in the practice of medicine] as now used could be sunk to the bottom of the sea, it would be all the better for mankind—and all the worse for the fishes." Around the same time, humoral medicine, particularly bloodletting, also fell under a bad light. The alarm bells were ringing.

Now add Upton Sinclair's revelations about contamination in the meatpacking industry and the result was a public screaming "do something!" in the face of the government. And it did by passing the 1906 Pure Food and Drug Act, creating the FDA. Meant to ensure the safety of food and drugs, the law also had the unintended consequence of leading to the creation of an industry around patentable drugs by modifying natural substances so they could be controlled and profitable—and if you think therefore other, more benign reasons read Gerald Posner's book *Pharma: Greed, Lies, and the Poisoning of America*, which shows how George Bernard Shaw's insights into the foibles of capitalist medicine profiting off our illnesses has been compounded by the pharmaceutical industry and its money.

The aim of the FDA was kind of like establishing a Department of Panacea. We know Panacea these days as meaning 'a remedy for all diseases', but most of us don't know her origins as the daughter of Asclepius who became the goddess of universal remedy. The FDA claimed the responsibility of assuring the public that their approval meant the drug was both safe and effective. They had to fit into Panacea's bag, but the effort was doomed from the start because fororphenlone, there is no panacea in medicine, and for two, the FDA excluded one of the other daughters of Asclepius, Hygeia. You may recognize that name as the root of the word *hygiene* and start to see where we are going with this idea. The FDA failed to include hygiene under its umbrella because the soap industry used its power to escape FDA control. It is great for me and my company because it left the door open for making Xlear. Its health benefits come from its cleaning the airway. But it's sad, too, because these two daughters of Asclepius—remedy and hygiene—were born to work together.

Dividing these sisters has hurt us all. The improvements in life expectancy seen in the 19th and early 20th Centuries was overwhelmingly due to hygiene. But hygiene was forgotten by the system after the FDA took up the role of making things better. They had a role also in making sure our foods were safe, but only saw that as making sure they were not contaminated. They left the efficacy to the food industry, which, we know now, led to sugar industry sponsored studies that got them off the hook for much of our more common problems.

Our food needed to be safe, but our drugs needed to be safe and effective. And of course our model for determining both is the analytical one that cuts the organism apart in our randomized, double-blind, placebo controlled crossover studies. If we instead realize the interconnections and complexity of living things, it would be apparent that the only way to measure both safety and efficacy is by how soon they die. That realization and its documentation in the medical literature was what ended bloodletting, what gave ORT a decided plus in treating cholera, and what continues to give the American Chemical Society a pain in the rear when their many byproducts are shown to shorten lives. Looking at the safety of new drugs through the lens of life expectancy, as described earlier, is a good idea.

Even if we had combined food with medicine and truly implemented the vision of health promoted by the ancient Greeks, we would have seen from the beginning that there is no panacea—no such thing as a remedy for everything. And maybe we would have seen that the analytical processes used in making a drug are not appropriate for the complex physiology and biochemistry of a living person. They ignore complex interconnections and miss the opportunity to prevent side effects, resulting in drug recalls that may not happen until after the damage has been done. Yes, the regulatory process for approving drugs is better than the nothing we had before 1906, but it falls short of what the public needs. It also creates a false sense of trust that removes the "buyer beware" caution that we should all use before putting anything into our bodies—by mouth, or by injection.

Despite this drawback, our system today exposes us to the FDA-approved medicines we call drugs to the extent that we no longer consider medical solutions beyond what is categorized as a drug by the FDA. Talk about cornering the market! There is room for a health food industry within this scheme, but regulations stifle its voice and prevent the manufacturers from making any bona fide drug claims. It robs the people of a proper understanding of how foods can help with their illnesses. This is clearly seen with oral rehydration, where a person selling, and advertising, ORT cannot state the *fact* that its use "has saved more lives in ten years than penicillin did in forty", because that is a drug claim.[39]

39 Editors, "Water, with Sugar and Salt," *Lancet* 2, no. 8084 (August 5, 1978): 300–301.

When I broke new ground with my xylitol nasal spray, I created a medicine with the potential to provide crucial prevention benefits during a pandemic, and I could not even get through the front door to talk with regulators. In a healthcare system run by drug companies, the entry fee to the regulatory process starts at $1 million if you want to eventually be able to say that the drug you created saves lives. That's only the entry fee—the final bill may be much higher.

I told the FDA I had an effective way to clean the nose, and when I told them the results of a clean nose they equated my food-based remedy to a drug and therefore subject to all the expensive hoop-jumping. Thanks to the soap industry's lobbying in 1906, the FDA does not control products advertised to clean the body. We were able to market xylitol in a nasal spray as "soap for the nose" as long as we made no clear medical claims.

Many ask: Why not make it into a drug? I filed an Investigational New Drug application with the FDA, but I did not have a pharmaceutical-grade bank account for their required studies. So I knocked on the doors of several pharmaceutical companies with my story, and one of my contacts went as far as to file, and be granted, a patent on my spray that was nearly identical to mine. But, after initial interest, they all declined because xylitol is not controllable from a regulatory perspective. People can make their own nasal spray and certainly *would* if the alternative were to pay pharmaceutical prices.

Joe Zabner, from the University of Iowa, tried harder. Interested in helping kids with cystic fibrosis, he pioneered the use of nebulized xylitol in a manner more amenable to control that might wind up being profitable. To provide a virtually unending series of laboratory and clinical studies demanded by the FDA, Zabner's group showed that:

- xylitol pulls water into the nose and enhances some of our innate defenses that are crippled by too much salt;[40]
- it is not absorbed through the nasal tissues but is delivered with the mucus to the stomach;[41]
- it is safe when inhaled in a nebulizer into the lungs;[42]
- it lasts about six hours in the airway.[43]
- But the FDA kept asking for more, and he finally gave up, too.

The fact that xylitol pulls water into the nose is significant. By itself it shows the ease of compensating for the lack of humidity in the air we breathe, but the amount he used was the normal concentration of solutions in the body. That's less than half the water-pulling power of the spray we used. Their research and our experience have proven the nasal use of xylitol to be a safe and effective way to counter the loss of adequate humidity needed for maintaining optimal cleaning processes. Optimizing our primary respiratory defense makes us healthier. The alternatives, as we have discussed, make us sicker. Common sense tells us which is

40 J. Zabner, M. P. Seiler, J. L. Launspach, P. H. Karp, W. R. Kearney, D. C. Look, J. J. Smith, and M. J. Welsh, "The Osmolyte Xylitol Reduces the Salt Concentration of Airway Surface Liquid and May Enhance Bacterial Killing," *Proceedings of the National Academy of Sciences of the United States of America* 97, no. 21 (October 10, 2000): 11614–19.

41 L. Durairaj, J. Launspach, J. L. Watt, T. R. Businga, J. N. Kline, P. S. Thorne, and J. Zabner, "Safety Assessment of Inhaled Xylitol in Mice and Healthy Volunteers," *Respiratory Research* 5 (September 16, 2004): 13.

42 L. Durairaj, J. Launspach, J. L. Watt, Z. Mohamad, J. Kline, and J. Zabner, "Safety Assessment of Inhaled Xylitol in Subjects with Cystic Fibrosis," *Journal of Cystic Fibrosis* 6, no. 1 (January 2007): 31–34.

43 L. Durairaj, S. Neelakantan, J. Launspach, J. L. Watt, M. M. Allaman, W. R. Kearney, P. Veng-Pedersen, and J. Zabner, "Bronchoscopic Assessment of Airway Retention Time of Aerosolized Xylitol," *Respiratory Research* 7 (February 16, 2006): 27.

better, but in our current regulatory environment, common sense has no chance of prevailing.

Respiratory Defenses and the Pandemic

Much of this story was told in Chapter 1 because of it's relation to our pandemic. The lab studies from Utah State University were confirmed by other institutions essentially showing that xylitol interfered with the adherence of the virus to the receptor binding domain used. These were not done on humans. A small study was done in Florida showing its efficacy in people with documented COVID that is referenced in the earlier section, but there is likely more data in anecdotal form in the comments section for Xlear® on the Amazon web site than there is in the medical literature. Absent in public communications about the pandemic is any mention of grapefruit seed extract as being viricidal, and even its close chemical relative, benzalkonium chloride, an FDA approved preservative, had been shown years ago to kill corona viruses.[44] Did that information cross President Trump's desk and lead him to comment about the propriety of using "disinfectants" to cope with the pandemic? While roundly criticized and ridiculed by the liberal media, his question would have been spot-on had he said preservatives rather than disinfectants—despite the fact that "disinfectants" is in the title of the article. It makes one wonder what power was behind the rabid media response. Benzalkonium chloride is used as a disinfectant and antiseptic as well as a preservative. The problem with using them to kill the virus is

44 H. F. Rabenau, G. Kampf, J. Cinatl, and H. W. Doerra, "Efficacy of Various Disinfectants against SARS Coronavirus," *Journal of Hospital Infection*, Oct. 2005; 61 (2): 107–11.

two-fold: it would lead to resistant mutations and you would have to use it more often—even possibly hourly in order to keep a functioning amount in the nose since the mucus is replaced every fifteen minutes.

Once again, though, the FDA stood in the way. GSE and related foods like limonene are not approved for use as antiviral drugs delivered via a nasal spray. To make that claim, the FDA said we would have to start from the beginning. ORT and the xylitol gum industry face the same high hurdles. Any food with drug effects is in the same "non-profitable" box, which is why no one is interested. Instructions for extracting GSE from grapefruit seeds are available online. See the problem?

My patent on the nasal use of xylitol expired in 2017, so if you can't find Xlear available for sale, you may wish to make it yourself or ask a compounding pharmacy to mix it for you. Here's how:

1. Start with a 45 ml (1.5 oz) bottle of saline nasal spray with GSE or benzalkonium chloride as a preservative. "Ocean" is the name brand I first used but generic equivalents are available.
2. Add 5 grams (a level teaspoon) of xylitol. Let it dissolve for a few hours.

It is also important to spray properly. The nasal passages don't go up, they go backward toward the back of the head, so aim the spray that direction. Tip the head forward. Sniff and spray simultaneously to make sure the liquid reaches the back of the nose. If it drains out after you spray you know you are not sniffing hard enough.

Amcyte Pharmaceuticals, an Argentinian-based company, makes a version of Xlear. They did a study at the Biocontainment

Laboratory at the University of Tennessee to look primarily at iota-carrageenan, a seaweed-derived substance shown to destroy the coronavirus like GSE.[45] The study also looked into the effect of xylitol. Seeing no further viral activity in the xylitol-only arm, they called xylitol viricidal, which is wrong because xylitol does not kill viruses or anything else. Instead, it just vastly outnumbers the cellular docking sites, so the viruses bind with the xylitol and are washed out.

I was not surprised by their findings—xylitol is the All-Pro nose tackle of the body's defensive line. It and other flexible sugar alcohols provide decoy docks for adherence with microbial lectins, the hands they have to hang on with.[46] Adherence is the first step in any infection. Blocking it, as Costerton suggested nearly fifty years ago, is a significant advantage for coping with infectious diseases. Our mucins are already on the job with their variety of glycans—giving them a few billion molecules of xylitol that also fit into the microbial lectin binding sites only helps more.[47] With their entrance, *into* our cells for viruses, or *onto* our cells for bacteria, blocked, microbes succumb to the pressure that Ewald demonstrated, adapting to circumstances by becoming more friendly to the host. If only we could teach Congress to adapt similarly!

Before the coronavirus pandemic a group of doctors in Florida were studying xylitol and chlorpheniramine, an antihistamine

---

45 Bansal S, Jonsson CB, Taylor SL, et al. "Iota-carrageenan and xylitol inhibit SARS-CoV-2 in cell culture." A preprint available at https://doi.org/10.1101/2020.08.19.225854. The information on xylitol is on page 10.

46 Ferreira AS, Silva-Paes-Leme AF, Raposo NR, da Silva SS. By passing microbial resistance: xylitol controls microorganisms growth by means of its anti-adherence property. Curr Pharm Biotechnol. 2015;16(1):35-42. doi: 10.2174/1389201015666141202104347. PMID: 25483720.

47 Michael R. Knowles and Richard C. Boucher, "Mucus Clearance as a Primary Innate Defense Mechanism for Mammalian Airways," *Journal of Clinical Investigation* 109, no. 5 (March 2002): 571–77.

shown in a Chinese study to block the flu virus.[48] When the pandemic hit they switched gears to study xylitol, GSE, and COVID-19.[49] They developed a spray with both chlorpheniramine and xylitol and looked deeper into xylitol and its use for treating patients with COVID-19. Their study of three patients verified as infected with the virus showed that using the spray four times a day significantly shortened the time and severity of illness.[50] The study was blinded with neither the patients nor the doctors knowing whether they got the active 'drug.' But when one patient did not get well rapidly he went to his drug store and bought a bottle of Xlear, then he got better too—.so much for the double-blind, placebo controlled study.

Unfortunately, regulators said that telling people about this study, in an attempt to help cope with the pandemic amounts to a drug claim, and we already know there is no getting over that hurdle to FDA approval—regulators had been in bed with the pharmaceutical industry for many decades at that point. Evolution shows us that incestuous relationships are not wise for humans; inbreeding exacerbates all problems, and it is just as dangerous to the healthcare industry's relationship with regulators as it is in reproductive relationships.

The incest began during the AIDS crisis: as those affected were crying out for life-saving drugs, the government enlisted the

48 Xu W, Xia S, Pu J, Wang Q, Li P, Lu L and Jiang S. "The Antihistamine Drugs Carbinoxamine Maleate and Chlorpheniramine Maleate Exhibit Potent Antiviral Activity Against a Broad Spectrum of Influenza Viruses." *Front. Microbiol.*, 2018; **9**:2643. doi: 10.3389/fmicb.2018.02643.

49 C. C. Go, K. Pandav, M. Sanchez-Gonzalez, and G. Ferrer, "Potential Role of Xylitol Plus Grapefruit Seed Extract Nasal Spray Solution in COVID-19: Case Series," Cureus 12, no. 11 (November 2020): e11315.

50 Soler E, de Mendoza A, Cuello V I, et al. (July 23, 2022) Intranasal Xylitol for the Treatment of COVID-19 in the Outpatient Setting: A Pilot Study. Cureus 14(7): e27182. doi:10.7759/cureus.27182

pharmaceutical industry's help for coping with the FDA's backlog of possibly helpful drugs. It led to the creation of the Prescription Drug User Fee Act, which authorized the FDA to collect fees from companies that produce certain drugs. In 1995, three years after the Act was signed into law, the Prescription Drug User Fee was $208,000. Twenty-three years later, in 2018, it had ballooned to $2,421,405.[51] It grew similarly as a percentage of the FDA's budget, from 2% in the beginning, to 26% or $628 million in 2009.[52]

Talk about a tactical decision that's not in our strategic interest! The incestuous offspring of this brother-sister relationship is unhealthy to say the least. The function of the FDA has essentially changed from regulating their industry partners to protecting them.

And the FDA is not acting alone. The Federal Trade Commission (FTC) protects their turf, too. In October 2021, the FTC sued Xlear for promoting an unsubstantiated treatment for the COVID-19 pandemic. The company's counterargument was that their claims come from studies at approved laboratories in the United States, confirmed by an independent lab in Switzerland, showing that the preservative GSE is toxic to the virus and that xylitol interferes with its adherence in the nose. Adding to their defense are the results of the small clinical study done in Florida. and the later larger one, referenced earlier, done in India's hospitals by researchers associated with the Wellcome Trust, the largest research funding organization in the world except for the National Institutes of Health. It was done to see if

51 Caroline Chen, "FDA Repays Industry by Rushing Risky Drugs to Market," ProPublica, June 26, 2018, propublica.org/article/fda-repays-industry-by-rushing-risky-drugs-to-market.

52 The Prescription Drug User Fee Act: History Through the 2007 PDUFA IV. Available online at: https://www.everycrsreport.com/files/20080627_RL33914_d23ea7f2e60b79167c69e2ff075a6d2d8441c534.html#_Ref224541062

COVID could be prevented by using their version of Xlear. but it was more of a sales pitch than a study because the agent studied was not described so no one could replicate their study. All they told was the ingredients—several osmotic agents, xylitol as the third ingredient, and a bunch of essential oils, like GSE, with known antibacterial effects. And so far their spray has not made it to any shelves.

There would be no issue if the agencies wanted to learn by sharing information, but their focus appears more to protect the pharmaceutical industry and its profits.

Bangladesh faced a similar situation during a 1980 cholera epidemic, but rather than quibble over regulations, a well-funded NGO there sent workers into the affected communities to tell them how to make a homemade version of ORT—the recipe that was provided earlier. It saved countless lives and put ORT on the map. Success is hard to deny when a million lives are saved. But like Xlear, ORT is not a drug, so there is no financial interest in promoting it. If you are a suffering patient it's a different story, but the system doesn't seem to care.

Just as ORT helps with cholera, Xlear is a safe, effective, and inexpensive way of coping with coronavirus, and keeping your respiratory defenses optimal should help with the flu as well; in the lab erythritol did a better job preventing the adherence of the flu virus than xylitol, but not nearly as well as chlorpheniramine. The compound spray of xylitol with chlorpheniramine is available in Mexico if you live close to the border. But the absence of research does hurt; it would be nice to know if the chlorpheniramine worked as an offense or a defense, i.e. does it kill and lead to resistance, or does it negotiate and lead to friendlier microbes. Flu season is like cold season for the same reasons discussed earlier: it's a respiratory infection and breathing dry air cripples airway defenses. Estimates of the cost of these pandemics and epidemics approach $6 trillion dollars in the U.S. alone and we are still adding to the bill. On the other hand, the cost of keeping our noses clean and free of this virus is around $60 dollars per person per year.

Princeton professor Fintan O'Toole says, "Change happens when the cost of continuity outweighs the risk of transformation."[53] In the U.S., the cost of continuity is another $1 trillion dollars; the cost to subsidize everyone keeping their noses clean is *6,000 times less*. The problem is that government regulations prohibit the spread of information, and stifle new research—and the NIH gets royalties from the vaccines.

The purpose of the Biomedical Advanced Research and Development Authority (BARDA) of the Department of Health and Human Services is to expedite treatments seen to affect disease processes, which includes non-pharmaceutical products needed to combat health security threats. Unfortunately, by the time the benefits of xylitol and grapefruit seed extract came to their attention Fauci had shifted their funding to HIV immunization.

Time and again, we knock at the door and come up empty.

Defense medicine is a paradigm shift, and I believe it is an easy one to make—the benefits are so clear and the savings enormous. The communities we live in extend beyond our fellow humans to include even the bacteria, both the pathogens and the friendly commensals that help and protect us. The common sense of a defensive approach to not just coronavirus but microbes in general is simple enough for a child to grasp, but the adults are

---

53 Fintan O'Toole. "To Hell with Unity." *New York Review of Books*, 25 March 2021. P.10.

too caught up in their war—a war they can never win, but that is so profitable it can't be stopped.

Paul Ewald, as discussed earlier, looked at how to bolster our defenses, methods we are all familiar with: clean water, masks, gloves, isolation, social distancing, hand washing, condoms, bed nets, etc. It worked like a charm when cleaning the water supply shifted the cholera bacillus to the *El Tor* strain that is not as lethal, and the Japanese use of condoms shifted the HIV virus to less virulent strains. How much more hard evidence do we need to see the wisdom of ending our war with microbes?

We all know the benefits and the methods that keep us safe. They force adaptation by microbes to be less deadly. And these defenses, both external to us as well as internal, do not threaten the microbes. Our offensives against them do threaten them and seem to only make the situation worse as they adapt by becoming resistant to our weapons. In turn, we develop new antibiotics in an arms race to see which side will develop the deadliest weapons. But unlike us humans, the microbes do not have a choice. They follow how evolution taught them to adapt. We, on the other hand, should know better. Does anyone ever win an arms race? What if a microbe mutates in such a way that it spreads easily, we can't stop it, and it's deadly to most people it infects? That's when the UN 2050 prediction becomes reality. The remaining humans alive on earth can then debate the merits of an offensive strategy.

If not for the loss of revenue faced by the drug monopoly, a paradigm shift should be easy. We have nothing to lose and everything to gain. Furthermore, the health and safety reasons that prompted us to create bureaucracies like the FDA need reviewing; the regulators have become part of the problem. It is clear that the warp of self-interest is far stronger than the woof of our health and well-being, and not just our own health but that of our communities. And we must accept that communities include all living things—even microbes.

We will now turn our attention there to see what we can learn from microbes and how they organize and interact as communities billions-strong. It offers a blueprint for how we can save ourselves by transforming our view of health as the promotion of well-being and reorganizing our society around it.

We are in trouble. The problems we face as a society are so bad and impossibly complex, no less than a complete change in how we think will save us, and it begins with what we can learn from microbes and how they organize. The war is over. We lost—and lost to something smaller than the eye can see. Now we turn to it to save us. It was here *billions* of years before us, after all. It knows a thing or two about how to survive on this planet. Using those principles, we have learned how to improve our own health. Now let's look at how to fix the system.

CHAPTER FOUR:

Fixing Health Care

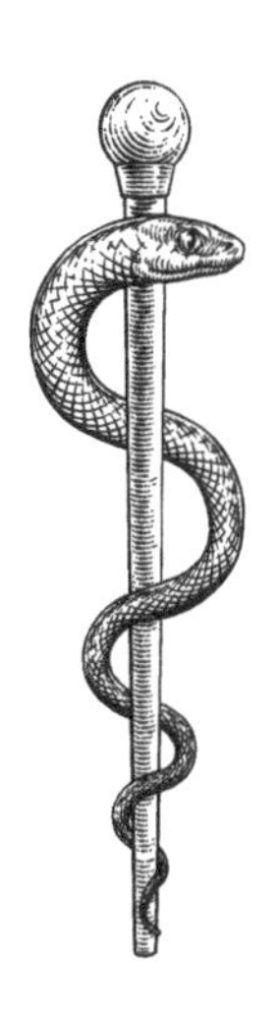

Changing the system begins with changing the culture, and that begins with us all realizing the value of seeing differently. The primary problem is its cost; the secondary problem is its orientation—and of course they are related.

If you printed out and stacked up everything written about what went wrong with the U.S. healthcare system and how to fix it, we might have to get NASA involved because the stack of paper would reach into outer space. And despite that endless output of analysis and expert opinion, the core issue behind the escalation of costs is rarely mentioned: patients were replaced by insurance companies as the primary payers of healthcare costs, destroying the balance of the marketplace. Monopoly is the goal of those working in the market; that's where one entity controls the particular market, like we saw in our past with Standard Oil and AT&T where they can charge whatever they want or their product. Health care is moving in that direction, as today a handful of top dogs control a healthcare system that delivers services to hundreds of millions of people. Their lobbyists and the power of their immense wealth has purchased them a free pass with their regulators, enabling them to charge more and deliver less. The companies say they want to keep costs down, but their primary focus is still making a profit, and without a watchful eye on what's paid for and why, they are free to game the system and run up the bill.

We, the patients, are their golden eggs, and the more they run us through their system, the richer they get. It's healthcare by the dollar. By squeezing chronically ill patients out of their marketplace and into Medicaid insurance companies become the primary payers, no one is on the ground to question the necessity of tests or procedures or medicines. No one shops around for better prices or alternatives. There's no real accountability. And the patients pick up the mistaken belief that the most expensive care is the best care. That leads to epic collusion and abuse and a healthcare system that's the most expensive in the world but delivers third-world care.

Linda Gorman, a PhD economist specializing in healthcare policy, concludes:[3]

> Two clear choices face those who would shape future U.S. health care policy. Continuing to follow old habits of layered regulation, third party payment, and increasing government control will continue the current cost spiral and the recent deterioration in patient care. To protect a bankrupt Medicare program, government involvement will be extended into every nook and cranny of U.S. medical care. The regulatory overload will end private medicine and encourage those who can afford it to purchase their health care abroad.
>
> The other choice is to deregulate, returning insurance to its traditional role as protecting against bankruptcy and promoting saving to pay for the higher health expenses that generally accompany old age. Let consumers spend their own money on health care, free of interference from professors with statistical studies and bureaucrats with specific notions of how people ought to behave. This is the choice that has the potential to stop the cost spiral, lower costs, and provide better health care for all Americans.

Gorman worked for a libertarian think tank, and for that reason is viewed suspiciously by some, but her quote above hits the nail on the head. There is no end in sight to the cost escalation, and without the patient as the primary payer, the only other choice in this scenario is for the government to regulate. And as Gorman predicts, it will be onerous and ineffective, as if it isn't already.

The market needs to regulate the industry through the choices individuals make based on the circumstances on the ground. Studies show that when patients are the primary payers for their health care services, they access the system less frequently because they do a better job of weighing costs versus benefits. You don't need a PhD in economics to know that people don't care as much about costs when someone else is paying the bill, but their minds would quickly change if they knew that they really are paying in other ways.

One big and ugly hidden cost is the decline of our healthcare system. When the patient was the consumer, the interest of medical professionals was to improve patient health, and it worked dramatically well when seen from a historical perspective. Data from the University of Oregon's History Department shows life expectancy moving from 38 to 68 during the century from 1850 to 1950, an average increase of three years per decade. Life expectancy is how we measure the efficacy of a nation's healthcare system, and there is no question that the dramatic increases in life expectancy in the U.S. were due in large part to public health measures like clean water, sanitation, hand washing, chlorination, and pasteurization. For the first century it improved by three years per decade. From 1950 to 2000, it increased by only 9 years—less than two years per decade. The question is why. While correlation is not causation a likely suspect is what insurance did to the orientation of healthcare in moving it from the patient to the profit.

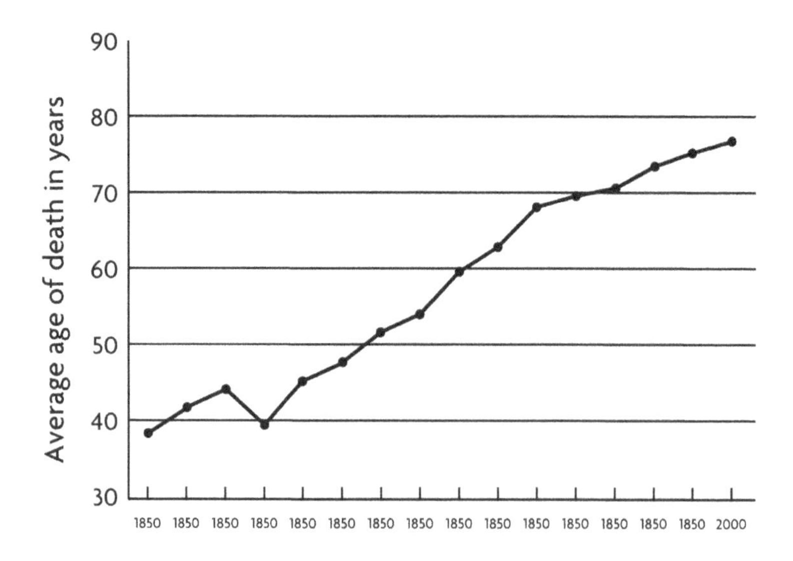

1950 was the period when health insurance invaded America's industries. Thanks to the post war wage freezes and health insurance being tax free, industries could entice the needed workers by adding health insurance–which included both hospitalization and office visits. The role of the consumer-patient as the brake on costs in the health care marketplace was eliminated—and profits took over. That's also the period when cold pills were first prescribed by doctors.

By 2015 the increase in life expectancy basically stopped. The opioid crisis is playing a part because it is driving up mortality rates among young people, but the Affordable Care Act furthered healthcare's commodification by moving more control to the insurance side of the industry. Doctors, frustrated with the expense and arguments with insurance companies, are moving to hospital clinics where profit interests demand both volume and efficiency over patient care.

So, there are basically two elements driving the relative, and now absolute, decline in our life expectancy, and it isn't, as some argue, that we ran out of new ideas for public health measures. First, I argue that the neo-humoral thinking behind the suppression of symptoms that are actually defenses is crippling the survival traits that we developed through natural selection. Second, third party payment through insurance and the resulting commodification of health care move the system to align with the bill payer and away from the patient's best interest. Both are driving the decline in the efficacy of the system. Health care had traditionally been a service provided by professionals. Third party payment pushed it into a market orientation to be codified and tracked down to the last tablet of aspirin; the goal in a market is to optimize profit—well being takes a back seat. Without the customer playing their role in controlling costs, the market grew sick from greed and shortsightedness. Shaw's profit oriented catastrophe is reality. The opioid crisis, driven by the callous profit-seeking of companies like Purdue Pharmaceuticals, with their regulatory approval, only made the problem worse.

It centralized billing and opened the way for more proficient gaming of the system. The industry blames the decreases in life expectancy on increases in chronic illnesses—hypertension, diabetes, obesity, kidney disease, and more—which is technically accurate. But one would think that the response of the industry would be to put more effort toward prevention, rather than the feasting on profits that are so obviously compromising the system. If the South African study showing the normalization of diabetic rats is accurate, for example, supporting xylitol production could reduce its marketplace price and encouraging its use in human diabetics would likely come up with the same results as in the rats. Instead, the effort is spent on developing medicines and procedures that manage symptoms. It's a bait-and-switch. Patients live longer with their diseases, and that, my friends, is the sweet spot in a profit-oriented healthcare system. Don't provide cures; provide medicines and procedures. Keep the profit wheel turning.

How We Got Here

Third party payment was not an unreasonable response to the need for expensive new medical technologies and life-saving surgeries, especially after the spike in demand for them with the development of anesthesia. Hospitals were needed to use this technology, and nurses were needed to care for the critical patients, and those are some big expenses.

Insurance entered the scene to meet the need, first in Texas. At the beginning of the Depression, Baylor Hospital began charging Dallas teachers a monthly fee of $0.50 for health care. In today's dollars that's $8.68. Compare that with the $1,000 dollars or more per month that people pay on average today and it tells the story of massive inflation in costs compared with the rest of the economy. Soon after Baylor's experiment, Blue Cross entered the market, providing insurance for hospitalization. It came in handy for emergencies. A major boost to insurance rolls came from automakers who needed employees but faced a postwar wage freeze, so they made the tactical decision to provide health insurance as a benefit. Insurance no longer paid just for expensive and emergent hospitalizations, it paid for health care, and since insurance was footing most of the cost, everyone could afford Cadillac care, giving us the mistaken idea that good health care is equated with its cost. It was cemented and expanded by the entrance of government, with Medicare and Medicaid, in 1965—also a tactical, rather than strategic decision.

Gorman's analysis saw behind the scenes at what was driving the healthcare system to change so fundamentally: with health insurance footing the bill, the patient no longer cared as much about costs. When patients paid the bills, it acted as a brake on cost increases. But costs ran out of control when insurance companies and the government stepped in. The AMA at the time blamed government intervention for the escalation of costs but remained blind to their involvement with the earlier insurance.

Employer-based health insurance is our accepted model, and it is part of the culture—most full-time jobs offer it. Patients want Cadillac care, and doctors are happy to comply. The insurance companies pass on the costs to employers at a markup and make their profits that way. Employers pass on the costs by inflating prices and economizing, which is just a fancy word for giving us less, another hidden cost. No wonder employers are moving overseas—it removes the expense of paying for health insurance, which began its life as a great idea to meet emerging needs, but like the baby hippo that grew up, it eventually ballooned into a behemoth that won't budge. It is simple economics, and it all went wrong when the patient no longer provided the market pressure to keep costs down.

As a doctor who knows the history of medicine, I can tell you that the self-centered, warp point of view of putting profits first is nothing new in medicine. But before health insurance fundamentally changed the game, the raw sort of selfish, profit-driven behavior we see today across the board was aberrant. Turning the service of health care into a commodity made medical practice a numbers game. Doctors were advised to set their prices toward the high end, since insurance was picking up the tab, with the goal of driving up the numbers. And it affects our decision-making.

When I worked in emergency medicine, I was encouraged to order an IV for anyone who looked like they could use it, which added a few hundred dollars to the bill. The IV itself was profitable, but the real money was made by the increased 'level of care' billed to insurance. For the provider it means "cha-ching!" I resisted this practice because in most cases a bottle or two of Pedialyte is sufficient to reverse the symptoms of dehydration. Pedialyte costs less than a dollar per bottle but using an IV instead racks up the patient's bill. With countless ways for frontline decision-makers to game the system, they add billions of dollars per year in extra billing.

Cha-ching!

The consequence of changing the paradigm to one of seeking profits by gaming the system and running up the bill, is that well-being is an afterthought.[2] The growth of third-party insurance was initially seen as a reasonable, tactical decision in response to the need to pay for expensive new procedures in health care. It's a market-based solution, and it's classic capitalism. But shortsightedness opened the door to the unintended consequence of creating more demand for healthcare services, and it created the mistaken perception that the cost of healthcare is a true reflection of its value. It is the reason for the chasm between what we pay for and what we get. We pay for Cadillac care but are lucky if we get Yugo care[54]. The decline in life expectancy is but one symptom of a sick system.

---

54 The Yugo, for those not familiar with this vehicle, was made in Yugoslavia back when it was a nation. It had an indestructible plastic body, but the rest of it did not merit much. A response at an Auto Parts store is not inappropriate: A man walks to the counter and asks if he can get a radiator cap for his Yugo. The sales person says: That seems like a fair trade.

Several decades of the money chase as the status quo in medicine have inbred this system, so it will not be easy to change. The industry shares a bed with its government regulators, and neither have an interest in fixing the empire they built around health care. The fix will have to come from the ground up, and cross-pollination with the root stock is the fix for inbreeding. That means the interests of the system need to be returned to the patient and their health. It all begins and ends with the consumer.

Can We Get Some Empathy, Please

Absent the empathy described in his earlier *Theory on Moral Sentiments* that was foundational for the marketplace of Adam Smith, economist Joseph Stiglitz argues for a transparent marketplace. Either way, something is needed to inform the consumer about what he is buying. According to Adam Smith, empathy is what drives self-interest to align with community interest and provide the "invisible hand" that theoretically guides mutually beneficial choices in market-based economies. The irony is not lost on me when I say that we need more empathy in the U.S. healthcare system.

When it comes to health care choices, the use of health savings accounts (HSAs) are an example of the invisible hand at work. Most people who can afford them are happy with their HSAs, but people with multiple or chronic health problems often run dry and preventive measures are not used in routine health care. Singapore's Central Provident Fund (CPF), described well by William Haseltine in his book *Affordable Excellence*, better manages those situations. The CPF is a retirement fund that may be used to pay for routine health care.[5] This fund, like our

Social Security, is paid for by the person and their employer and run by the government. But unlike Social Security, the CPF is controlled by the individual. And it can be used for other socially beneficial needs like health care, housing, and education, and it can be shared within the family, which mitigates some of the slow build-up. Like HSAs, it changes the way the citizens in Singapore make decisions about their health care, and in Singapore there are also other ways to cope better with chronic illnesses.

Their system changes the way they're treated and as a result, they enjoy some of the best health care in the world, and the cost of it is shockingly reasonable. Bloomberg ranks it as one of the best in he world. It's what happens when you give power to the people and enable consumer choices to shape markets.

Social welfare is an important aspect of Commonsense Medicine because of its big impact on individual and collective health. Innovative solutions like the CPF can get us to where we need to be, and they are market-based and controlled by individuals, not by bureaucracies and corporations. I will reveal more of these solutions in a future book. Our focus for now is on healthcare.

Like a CPF, a "people's needs fund" (PNF), an update for HSAs, would solve the problem of needing to build up funds before HSAs can pay for major medical emergencies. PNFs, like CPFs are shared within families and every child is given one at birth by a government grant. As well as family sharing, community sharing to a lesser degree is also realistic. Many communities come together when someone is confronted with an overwhelming expense. This includes places of employment, churches, and any other community organization. Charity is best when it's local, and this avenue would strengthen all organizations involved as well as the community as a whole. It would be a powerful force in a good direction.

The reason it's so powerful is that it takes us back to Antonovsky's coherence. All three of his elements are promoted by community. Meaning comes from sharing stories with others; understanding often comes from others, and; all involved have and share in the control. The resulting coherence further helps the chronically ill; as well as lessening lots of the financial stresses that divide our families. Something like this is the best means currently available to us for empowering people. It would also enable us to start returning health insurance to its original role of protecting against catastrophic financial loss. That would shore up the system and reduce medical bankruptcies. These pathways are open to us, and they are the sorts of simple solutions to complex problems that work from the bottom up and put to use both the principles of market economics, and those building wellbeing. By replicating and building on the successful strategies of Singapore's healthcare system, which Bloomberg ranks as the world's most efficient, we could save nearly $2 trillion dollars a year, and be much healthier and more resilient.[55]

Singapore's system operates in accordance with the view that a healthy population leads to a healthy society. It does not necessarily mean that the government should pay for everything, or that medical care is a universal right, as Pope Francis and others have argued, though in principle I would agree. There are far too many harmful behaviors where the patient is clearly responsible: use and abuse of tobacco, alcohol, and other drugs, lack of exercise,

55 Singaporeans spend about 5 percent of their GDP on health care with 2.2 percent coming from government. We spend 17.9 percent, for a difference of 10.7 percent from Singapore's. Our GDP is now about $17 trillion annually and 10.7 percent of that is $1.82 trillion.

dietary excesses and more that adversely affect one's health. And it doesn't serve anyone well in the long run to let people off the hook for destroying their own health. Instead, because a PNF is a retirement fund that can be used as an emergency source of money to pay for health services, it applies pressure to make good health choices—it's *your* money, and it's in *your* interest to save as much as possible. It leads to enhanced health by encouraging healthy behaviors *and* making them in the person's financial interests.

The above suggestions address the underlying problems that crippled our healthcare system and skyrocketed its costs, and they tap into the sort of thinking that produces innovative disruptions. And they show more empathy for the individuals, families, and communities who are not being served well by the dysfunctional U.S. system.

Survival of Healthcare

As a physician, my first interest is the health of my patients, with health defined as their sense of well-being rather than just the absence of illness as defined by the medical profession. But well-being encompasses more than that. Health is best measured by life expectancy, and it improves when the patient is the focus of the system. It decreases when profits rule the system.

Our cosmic war with microbes is misguided. We know now that we can negotiate with our microbes, encouraging them to adapt in friendly ways and to live with us. We also know that our own health is directly related to the diversity of our commensal microbes.

Defense medicine is a paradigm shift in how we see health. When a person presents with bothersome symptoms our first question should not be how to balance them, but what do the symptoms mean for the body. Backup defenses are bothersome; they are a signal that our primary defenses are exhausted, and our body needs help. The exhausted defenses should be optimized as explained herein. The symptoms should not just be turned off through the use of drugs.

The use of PNFs to pay for routine care restores the patient to the marketplace of healthcare. It may be seen as more expensive, but in eliminating the intermediate insurance company with their profits, and reestablishing control by the patient the overall price will be less. That's what Linda Gorman's experiment demonstrated. And it won't adversely affect our health.

These are four common sense ways to improve our personal health and that of our nation. They are strategic; they look at the root of our problems. All four of them build the control, understanding and meaning that make up coherence—with potential savings in the trillions.

But coherence too is on multiple levels; individuals, groups, and even nations have constructed egos and the coherence that accompanies them. I have tried in this book to discern between our constructed egos and the reality of who we are as a family of human beings. Egos, personal and national, often get in the way. We think mostly on an egocentric basis; that's what Social Darwinism teaches, but our health is more than we think. And if the Thomas Rule works in healthcare even *what* we think plays a role. Optimal health is not egocentric coherence; it's not the narcissistic thriving when the self is unknown, it's built with the coherence and the *meaning* that comes from belonging to the human family.

Promoting this type of coherence on a national scale requires a partnership between government and the citizens that helps them learn to use their problem-solving cortices to guide their fearful midbrains—ever alert to danger—onto paths that minimize threats. Or better yet, since most threats today come from *others* on the same path, maintain control of the midbrain by negotiating with the *other* to reduce counterwill by determining the best path forward. Learning this type of control is a major challenge; it requires an understanding of how to manage threats that is best taught in groups, which add meaning, and provide enough support that we don't revert to lizard brain thinking when confronted with a threat.

Productive change is not going to come from the system, nor is it likely to come from politicians in debt to their lobbyists, although PNF's would be attractive to both conservatives and liberals. Individual states could pilot such programs if they could convince their citizens of its value.

The easiest and most effective method is to work through combining the VA and Public Health Service, and to make more of our public eligible for their services. Both those organization have the health of those they serve as primary—they are not profit oriented. That change would make the others easier. All the others are seen as innovative disruptions in our current system. First among them is empowering patients with a sense of control by establishing government supported PNFs that are needed to escape from learned helplessness and put the people back into the healthcare marketplace. Given the lobbying power of the industry, this would best be done gradually, first by making PNFs available to those wanting them and gradually including more by enhancing their appeal. Expanding the PNFs to Medicaid and Medicare could only be done when they have been shown to work.

Second is the problem of the growth of the dealers in health care, as opposed to the healers—the same problem we define as hypocritic vs. Hippocratic medicine. Our culture of healthcare based on third-party payers builds the dealers into the system. The dealers survive because the patient has learned to be helpless. Giving the patient more control is the way out of helplessness and the way to push dealers to be healers.

Addressing both problems will help cope with the third: the cost. Growing at almost double the rate of inflation, it is a real threat to our future. A major reason for this is the industries' focus on profits.

During the coronavirus pandemic, for example, the focus of the industry, the government, and the media was on vaccines that help us get rid of the virus. That's also the most expensive way. Warfare is always more expensive than negotiation. The government went into a partnership with the pharmaceutical manufacturers and handed out billions of dollars. Money talks and it goes where those in power put it: in this case to building a consensus on the side of the pharmaceutical industry. But it's peanuts in comparison to the persistent need to profit from the system.

There is also little interest in our system in preventing illness. It was our surgeon general, the head of our Public Health, who had nothing to do with our medical practices or hospitals—or their profits—who blew the whistle on cigarette smoking.

We now have an epidemic of type 2 diabetes. But there is no interest, because there is no profit, in repeating *in humans* the experiments done on diabetic rats showing that eating xylitol rather than sugar prevented their diabetes.[56]

56 Chukwuma CI, Islam S. XYLITOL IMPROVES ANTI-OXIDATIVE DEFENSE SYSTEM IN SERUM, LIVER, HEART, KIDNEY AND PANCREAS OF NORMAL AND TYPE 2 DIABETES MODEL OF RATS. *Acta Pol Pharm*. 2017 May;74(3):817-826. PMID: 29513951.

Then, of course, there is COVID, and what we have described as an inexpensive method of taming it. But it's not just me. A Chinese researcher, without knowing of Costerton's earlier similar argument, has now written, "I think many glycans, including xylitol, might be able to interact with or interfere with the glycoproteins on the virus or cell surface [and] thus reduce viral infection."[57]

The resolution to this profit orientation is in pushing the non-profit patient-centered sector of the industry to do these studies. That's the combination of the VA and PHS proposed in the last chapter. These doctors and researchers are salaried, so their interest is in not working, and in order to do that they have to help their patients live independently without relying on the system. Defense Medicine and Hippocratic drugs (foods that are drugs) are fertile fields for clinical research and preventive medicine for researchers with that interest.

That's a bit better than how we can make an immunization for COVID by modifying our messenger RNA to make an antigen like the protein stalk of the virus—with the unstated and perhaps unknowing that this stalk segment is foreign to us yet it too is covered with our glycans as it exits our cells into our bodies. It's something that warp speed ignored. Half of our population does not trust genetically modified foods, but they line up for immunizations—absent the informed consent usually required explaining the side effects of becoming a genetically modified human—without an argument.

Proteins can be modified, controlled, patented and therefore profitable enough to be made into drugs. That's the warp,

57 Personal communication with Hongliang Wang in reference to his report: X Zhao, H Chen, H Wang. "Glycans of SARS-CoV-2 spike protein in virus infection and antibody production." *Frontiers in Microbial Biosciences,* 8,629873, 2021.

profitable, approach that continues our unwinnable war with microbes. But beyond the profit motives these vaccines are injected far from their normal entry into our bodies, so their actions are also different.

When a microbe enters our body, it runs into many parts of our immune system. When our defenses are optimal this happens before it finds a way to hang on to us. These immune system elements begin the program of building immunity, and it involves more than just one segment of our immune system. The COVID-19 immunization builds, for example, IgM and the T cells that supposedly build cellular immunity. "Supposedly," because cellular immunity is long lasting, and the immunizations have not demonstrated that. When introduced nasally the major immunity is via IgA as well as IgM, and the cellular immunity is built by exposure to the lymphoid tissue of the adenoids, which is absent with the vaccine going into your arm.

The fundamental problem with this is that immunizations should try as much as possible to replicate the pathway of natural immunity. That means they should be introduced to us in the same manner as done naturally: airborne viruses should be attenuated—made inactive—and delivered by nasal spray; ones delivered by contact, like smallpox should be by scratching the skin, etc. That's why the smallpox and Salk vaccines were so effective; both deliver a weakened virus that doesn't cause either smallpox or polio via the same pathway the real pathogen uses. In the case of COVID the proteins being studied have little to do with immunity.

All these proteins are called glycoproteins. The virus that causes COVID, after it has found a docking place, uses our own coating process to add our glycans, and once disguised, it can go anywhere in the body and cause all kinds of problems, and it can escape detection for years, as seen with "long COVID."

Osteopathic physicians in the 1918 pandemic had a mortality rate 40 times less than the rest of the healthcare profession. There are alternative and integrative physicians today that treat COVID early and prevent its spread throughout the body that experience similar benefits. There are not many examples of this, mostly because the old drugs repurposed for this have not passed FDA efficacy for that specific use, so those using them are condemned by the regulators—much to the agreement of their profit oriented pharmaceutical controllers. And the media overlook the financial interests of the system—both pharmaceutical and governmental, that experts on valid arguments agree weigh heavily on reported outcome—and repeat the message.

If we really want to avoid the next pandemic we need a way to avoid submitting to a collective that profits from them. Many in our third estate—those monitoring our governmental bodies and globally oriented industries—need a wake up call.

No one is trying to find reason behind why some people seem immune, or why even a saline nasal spray helps reduce the problem. There are simple ways to deal with pandemics that we have explained—even to the point of how they work. My son that runs Xlear Inc. is just one example. He called the other day to tell me that he tested positive for COVID antibodies. That usually means he either had the disease or was immunized. But he had had neither the disease nor the vaccine. He had been exposed several months earlier when sharing a houseboat for a week with someone who tested positive shortly after the trip, but he never got sick. He credits keeping his nose clean. Xylitol does that, not by killing the virus, but by giving it something it wants to hold on to—by negotiating in the cooperative pathway. That's the way our immune systems should work. Exposure to a microbe triggers an immune response whether or not the microbe is able to attach to us. Repeated small exposures like that build more immunity; it's how we learn to live with the microbes as many are now encouraging—and you don't have to get sick to do it. One of Trump's big lies, that the virus was nothing to worry about, was much closer to the truth than his opponents were willing to see; their blinders were set by the warfare, and the money is on their side.

Homeostasis is our dominant paradigm for the way we see our physiology and how we base our treatments. But *heterostasis* is more friendly to defense medicine. It was begun by Hans Selye, the Canadian doctor who studied stress and its effect on the human body, and he credited Claude Bernard for many of his ideas.[58] Heterostasis recognizes that the body needs to be out of balance sometimes; *hetero* means different. Stress affects what Bernard called the "milieu interieur", the internal balance of the body—what was later called homeostasis. Selye wrote about the challenge state, when the body recognizes a problem, the resistant state, when it tries to cope with it, and the exhaustion state, if it fails. The first two need help, optimal defenses and reduced stress; the third is when we feed the system

The other side of that is when we win. That's when our defenses are optimal. In cases like this our physiologic defenses are front and center for maintaining our health, even more than coherence. In my case, and hopefully yours, it was the control and understanding that defense medicine introduced that opened the door to coherence and salutogenesis.

---

58 Hans Selye. "Homeostasis and heterostasis." Perspectives of Biological Medicine, Spring 1973; 16(3):441-5.

That same resistance of self-interest and profiteering is what Colin Campbell found impacting our food processing industries. It's the same pattern behind all financial crises. Riskier options can pay higher rewards so self-interest opts for the profit. It's very much like 'the Prisoner's Dilemma,' a game used by psychologists to study the balance of self-interest and cooperation. In the original game a couple of crooks are apprehended by the police and questioned separately. Each of them has the choice of ratting on the other or not. Ratting (called defecting) can reduce your sentence and is in one's self interest since you get a great benefit. Not ratting is the cooperative avenue. Staying mum benefits both prisoners because the information is less, and so (likely) is the sentence. So both see a benefit, which totals more than what the defector gets. Computer models of this game have been played enough times to show that defectors eventually wind up with he least fit society; in a mixed society the defectors have the greatest fitness, and; a cooperative society has the greatest fitness. Building a future based on self-interest and survival is also risky because self interest leads to inequality—that's the result of defectors profiting off others that makes them more fit in a mixed society—but inequality poisons social trust, and trust is he foundation of a cooperative society. Such economies crash regularly; it's the small and diverse economy—the cooperative one—that is more flexible and resilient. In error we see the desirable and sought-after elements for building greatness as those measured by the GDP. They are not the elements, as Robert Kennedy notably said at the University of Kansas in 1968, that "make us proud to be Americans", nor are they the ones that lead to stability or resilience. They are important only for those focusing on the size and power of our economy. Well-being is a matter of democracy and societal choices. It's a matter of balancing the properly focused warp with the woof.

Your Health, Your Choice

In his book, also mentioned earlier *The Innovator's Prescription: A Disruptive Solution for Healthcare*, Harvard Professor Clayton Christenson suggests that we relocate simpler medical procedures from expensive hospitals to less expensive clinics and surgical centers. It's a good idea, but it's a baby step, a tactical maneuver that addresses a cost factor. And it's the sort of thinking that falls short of doing more than tinkering with the current model. We need a strategy—a big, bold one—and it needs to be a public health push toward cost-effective preventive medicine that focuses on what individuals can do to promote their own health.

Central to our strategy is the innovative disruption of defense medicine, with its focus on supporting physiological defenses. Natural selection gave us these defenses to inherit because of their benefits for survival, and we have wrong-headedly defined them as illnesses and created a variety of drugs that cripple them. Supporting the body's natural defenses rather than crippling them is the definition of 'common sense medicine', and is something we all need to do. Indoctrinated in the current paradigm and unable to see beyond their own interests, especially their profit interests, the "experts" give natural defenses little attention. Therefore, it is up to us to assume our own control and be responsible for our own health.

There is no better innovative disruption than boosting our health by optimizing natural defenses. The approach is simple enough for the public to understand. Christensen points out that the rise of alternative medicine is a reflexive response to the common and accepted practices that utilize expensive technological medicine. Alternative medicine prefers the use of the medicinal foods we call Hippocratic drugs that are relatively inexpensive. People are increasingly realizing that measuring care by its cost is not rational, and they are looking for alternatives. The door is opening, and the strategic changes introduced here can blow it wide open.

I am well aware that I am drawing an unkind parallel between a system focused on profit-making and the one I propose that focuses on natural health. I know many doctors who are as frustrated as I am with our system. Coping with the COVID pandemic brought many more of them out into the open. But most of my colleagues know nothing of defense medicine, and their frustration is not enough to fight against the goliath of a profitable system that includes both the government and the healthcare industry. The only way to win this battle is by taking it to the people.

And the people need a sense of control; they should not feel used, told falsehoods, or feel cheated. In a nod to Adam Smith, the economist who popularized the woof that markets of empathetic individuals, his markets were based on players being cognizant of and honoring their partners in a transaction. Such a market can self-regulate and lead to progress. The warp of the marketplace is seen in how it has lifted the standard of living for millions of people, but the unintended consequence of a *laissez faire* market is the unleashing of the self-interest of Social Dawinism, which destroys the trust needed for its progression. The innovative disruption we are proposing puts people back into the role of controlling the healthcare marketplace through their decisions about health care. When people are removed from the decision-making process, their loss of control leads to learned helplessness, especially when the person (i.e. the doctor or bureaucrat) in control is seen as the expert who can't be questioned. The patient is then made into a passive recipient, and worse, a commodity to be run through the medical mill. In my medical practice, the patients who took an active interest in their health clearly enjoyed better outcomes, in contrast to those who put themselves at the mercy of experts and the system. This block to participation in their own care, aside from removing their sense of control and coherence, subjects dependent patients to the third-leading cause of death in the United States—the system.[4]

The Health Savings Accounts (HSAs) and PNFs already discussed are good ways of returning people to a role in the marketplace where they have some *control* and can escape learned helplessness. Singapore goes a step further by running a public health system that competes with their private, profit-oriented hospitals; it's the option for many with chronic illness. In the U.S., we don't have anything like that except for VA hospitals, and despite the political campaign against the VA, if you ask patients experienced with both the VA and private hospitals, you quickly find out which they prefer. There has never been a bankruptcy that resulted from a VA bill. If we want solutions that work, combining the VA with the Public Health Service is a bold move; put the two government agencies most interested in making their populations healthier in the same boat with adequate funding and see what happens.

"But it's socialized medicine!" you will hear the industry-aligned critics cry out as they polish their golden eggs. But it's merely a cover for their fear that a better option would easily win a competition for the public's business. I tried to find a better alternative, but I always return to the view of George Bernard Shaw, mentioned in an earlier chapter, that sees a big problem for the health professions when illness is more profitable than health. The problem is baked into the system, and nothing short of changing the system will fix the problem.

As comic Chris Rock said about the reason why we will never cure AIDS: "There ain't no money in a cure, the money's in the medicine. That's how you get paid: on the comeback." As brutal as that assessment is, it's accurate. Cure a disease and you lose your customer base. Incentivizing health over profit is the only way to fix our market-based model for healthcare. After fifty years of understanding the dental benefits of oral xylitol it's clear that you can tell a profit-oriented dentist from one interested in prevention by what they tell you about xylitol. Nasal xylitol has only had twenty years, but one can tell the same already, and most of the industry goes with the profit. Incentivizing is going to have to start with the people.

Critics argue that taking the profit out of medicine results in less innovation. Why would anyone spend millions of dollars to invent the MRI scanner, for example, unless they thought they could profit from it? It's a good argument. The drive to make a buck and get ahead is powerful—Nietzsche wrote a whole book about it in *The Will to Power*. But it's all about ego, and evolution is a better option; it's about sharing the benefits of innovations and superior adaptations with the community. It's how they spread and become part of society and how the individual thrives because their community thrives. It's all about combining the warp and the woof to insure a healthy fabric.

In our capitalist system, everyone wants to optimize their income. Those providing a service need to profit from the service, and the commodification of healthcare has paved that road. VA doctors are salaried—unlike their colleagues in the profit-driven system, their income is not dependent on having enough sick patients to bill. They may not make as much money as their peers in the private system, but their incentive may be to have more time to spend on their families or their leisure, and it's a trade-off they are happy to make. To have more time, they need healthier patients. Why? Because healthier patients need fewer office visits, and fewer visits frees up more time for the doctors, incentivizing them to make their patients healthier and less dependent. And from my own investigation of the VA's clinics, prevention and healthy practices are front and center at every visit. It's another way we can capitalize on market-based principles to fix healthcare.

A challenge for the combined VA-PHS is to establish and coordinate groups of people with the same medical problems. Such groups provide all the elements of coherence, and they could be based online. They should be limited in size, though, to make sure everyone has a voice. The group provides meaning from other's stories, understanding by learning about their condition, and control because the focus of such groups is on what the person can do to make their life better.

Building social capital includes efforts to improve understanding and individual control. The efforts of the government are more often on maintaining the control of the experts. But their experts are more often those buried in their profitable paradigmatic foxholes that resist any threatening novelty. Our nation's response to COVID provides a very good example. Building social capital is ignored when hiding product information and sensible management tools that optimize our defenses, such as those discussed here.

Eligibility for VA care comes through military service; eligibility for VA-PHS comes similarly via a period of universal service to country after high school. Proposed in our sequel as a means to 'form a more perfect union' the result is eligibility for a public health option in healthcare that would gradually include everyone in a system aligned on prevention and health. This option opens doors (*control*) for patients to learn about their conditions (*understanding*) and join forces with others in the same boat (*meaning*). These elements of coherence are at the core of defense medicine.

HSAs tend to be harder on the poor and the sick—many people don't have the extra income to sock away—which is the basis of the liberal argument against them, but PNFs mitigate those problems by incorporating measures of family and community sharing. They empower people with greater control over their health care by giving them control of their own money. This doesn't eliminate insurance, as Gorman argues; it just returns it to its traditional role of protecting us from unexpected and expensive problems. HSAs are associated with catastrophic insurance policies to prevent bankruptcies; PNFs do the same.

Ending the Drug War to Make Us Healthier and Wealthier

In all of the foregoing the system holding the empty bag is the healthcare system. That's both unfair and unnecessary. The biggest obstacle we face to changing the healthcare system is in how to replace at least some of the money that will be shifted away from it. The industries making billions of dollars per year are not going to just stand aside. Fortunately, there's a win-win solution: end the war on drugs and move that money into the healthcare system.

Illegal drugs dominate the informal economy in the United States, with an estimated value of $10 billion annually, and more when times are tough.[59] Other estimate put its value as much higher.[60] All of the money spent on the demand side originates inside the country, but most goes to the supply side and leaves the country, ending up elsewhere. The government's answer was a declaration of war, a tactical decision that lacked any strategy other than winning at any cost. In fact, John Ehrlichman, Richard Nixon's aide on domestic affairs who, like Michael Cohen, took the rap and the jail time that should have been his boss's, told the reporter Dan Baum that the drug problem had nothing to do with it. As Baum reports:[61]

> "You want to know what this was really all about?" he asked with the bluntness of a man who, after public disgrace and a stretch in federal prison, had little left to protect. "The Nixon campaign in 1968, and the Nixon White House after that, had two enemies: the antiwar left and black people. You understand what I'm saying? We knew we couldn't make it illegal to be either against the war or black, but by getting the public to associate the hippies with marijuana and blacks with heroin, and then criminalizing both heavily, we could disrupt those communities. We could arrest their leaders, raid their homes, break up their meetings, and vilify them night

59 https://obamawhitehouse.archives.gov/blog/2014/03/07/how-much-do-americans-really-spend-drugs-each-year

60 Taylor Barnes, "America's 'Shadow Economy' Is Bigger than You Think—and Growing," *Christian Science Monitor*, November 12, 2009.

61 Dan Baum, "Legalize it All." *Harper's*, April 2916. https://harpers.org/archive/2016/04/legalize-it-all/

> after night on the evening news. Did we know we were lying about the drugs? Of course we did."

Since that war began it has cost over a trillion dollars and adversely affected countless lives, both here and to our southern neighbors where we have exported it—and where it has led to the cartel related violence and state corruption that leads to the invasion of refugees on our southern border. Those costs are not included. This war is unilateral, there is no state opposition. Very much like our war on microbes the other side is depicted as evil so there is little sympathy, but in both wars the opposition is certainly not weak and defenseless. And in both negotiating with the opposition is reasonable, but in the drug war the other party should be the users and not the cartels.

The annual cost of fighting the war is estimated at one hundred billion. The annual loss to the cartels is 10 billion. That money could be put to use to replace the profits that the healthcare industry stands to lose when the system is changed.

Baum's argument to "Legalize it all" ignores the dangers behind these drugs—they are not innocuous. We learned from Prohibition that the way to deal with black markets was to legalize them, so alcohol is legal now and, according to the surgeon general its abuse cost 249 billion annually. That's more than the costs of drug abuse. Legalizing drugs would reproduce those sad results. Making them legal, but under a doctor's care would provide safe places for those wanting them, opportunities to talk about options, which include safer drugs, rehabilitation, or even different drugs.

The drug war is also hindering the therapeutic use of mind-expanding psychotropic drugs, made illegal because people can have bad experiences and do real damage to themselves and others.

This problem would be minimized by medical control. Psychedelic drugs do have good uses, one of which is they help people with PTSD. Johns Hopkins established a center for such research that is looking at the uses of psilocybin, the active ingredient in magic mushrooms, for: stopping smoking, alleviating depression and PTSD, and gaining personal insights.

One example of how psychedelics help people gain personal insights is demonstrated by their use on a group of religious professionals.[62] It's a flashback to 1962 when such studies were legal. Back then, Harvard Professor Timothy Leary and one of his doctoral students gave doses of psilocybin to a group of willing theology students prior to their attending Good Friday services, desiring to see how many of them would have a meaningful religious experience. They all did. Almost thirty years later, psychedelic researcher Rick Doblin wanted to find out how these people were affected in the long run.[63]

> Each of his respondents felt that the experience had significantly affected his life in a positive way and expressed appreciation for having participated in the experiment. Most of the effects discussed in the long-term follow-up interviews centered around enhanced appreciation of life and of nature, deepened sense of joy, deepened commitment to the Christian ministry or to whatever other vocations the subjects chose, enhanced appreciation of unusual experiences and emotions, increased tolerance of

62 Johns Hopkins, Center for Psychedelic and Consciousness Research. See the list of studies at: https://hopkinspsychedelic.org/index-1.

63 Doblin R. "Pahnke's, 'Good Friday Experiment': A Long-Term Follow-Up and Methodological Critique," Journal of Transpersonal Psychology 23, no. 1 (1991): 14.

> other religious systems, deepened equanimity in the face of difficult life crises, and greater solidarity and identification with foreign peoples, minorities, women and nature.

I chose this example not to show how good psychedelic drugs are. Too many people have had bad trips on these drugs, which is why they are illegal. I chose it to show that they can be useful. They should be controlled because their benefits are better assured through their therapeutic use. When used as part of a therapeutic process, psychedelic drugs and therapy work synergistically. We can make the argument for legalization on those grounds, but the precedent was set long ago by legalizing alcohol, a potentially dangerous drug. It makes no sense for a mind-impairing drug like alcohol to be legal but mind-enhancing drugs are not. Done in a controlled, caring, and therapeutic environment, the option of a mind-expanding trip that is both introspective and self-evaluative seems far more attractive than addiction to mind-numbing alcohol or narcotics.

Therapeutic psychedelics could easily become the drug of choice for people escaping traumatic events with alcohol or narcotics, or even for those wishing to have experiences like the Good Friday theology students.

It is painfully obvious that solving the drug problem isn't the goal, because if it were, we would be putting the bulk of the money into addressing the underlying causes. The Organization of American States (OAS) tells us—repeatedly—that the greatest underlying cause is the willingness of people to buy drugs illegally. The demand is huge. Reduce the demand and you reduce the supply. Reduce both and all sorts of expensive problems go away. The OAS says that putting health and community safety first requires a fundamental reorientation of policy priorities and resources, away from punitive enforcement and toward proven health and social interventions. I couldn't agree more. Talking about the problems doesn't help, either. As the southern neighbors of the U.S. know all-too well, trying to talk somebody into thinking differently about a profitable drug war is difficult. They have borne the brunt of it, and now they are letting their people vote with their feet. Fleeing the violence in their countries caused by drug cartels creates immigration problems in this one. No wonder we have problems at the border. Maybe it would be a good idea to stop creating them in the first place.

The connection between our ongoing immigration problems and illicit drugs demands a strategic response, and understanding the lies behind the policies leads to a cultural change as well. The immigrants on our southern border are coming north seeking asylum from the violence and other problems in their home countries. Not said, but clear if you look, is the connection between the war on drugs and the problems it exports to those countries. The U.S. goes south and destroys the crops the peasants grow to satisfy American demand, such as cannabis and coca, which has only made the crops more expensive and profitable to grow, exacerbating the problems.[64] The immigration problems at the southern border are sure to continue until we end the war on drugs.

The other side of the coin is the economics of how the war on drugs shapes the informal economy of the black market. This is a swamp, every bit as nauseous as our government swamp, and draining it requires the same strategy—get the money out. The

64 Louisa Farah Shwartzman. "Trump, Biden ignore how war on drugs fuels violence in Latin America." UPI, 29 Oct 2020. See https://upi.com/7050938.

public wants the drugs, and the only way to get the money out is to move distribution and sale over to the formal economy. We should have learned that lesson with Prohibition—if the demand is there and enough money is to be made, the black market will supply what people want. We also have the lessons gained from countries like Portugal that decriminalized drug possession and educated their public, largely eradicating the social problems caused by the sale and use of illegal drugs. Plus, it's more compassionate to treat addicts like human beings burdened with a medical problem, rather than criminals to be locked up and forgotten. But that is apparently what Nixon wanted.

A baby step in that direction is to revise drug laws to lessen the penalties, as we have done with cannabis, but the big leap we must make is to legalize recreational drugs and allow for their use under medical supervision. Medical supervision would lessen abuse, eliminate overdosing, and provide opportunities to present better options like treatment and counseling. And it would pull those billions of dollars into the formal economy that otherwise disappears into the black hole of the drug war. That boost would also balance, to some degree, the costs to our healthcare system of accepting defense medicine.

A similar logic also fits our approach to opioids. They have enough usefulness that they are controlled, not outlawed. In Rwanda, morphine is delivered free daily to those in chronic pain, like many of the survivors of the genocide that swept over that nation in 1994. It prevents many problems and is a compassionate response that builds goodwill between the people and their government. In the U.S., we have an opioid epidemic driven by an industry that got rich pushing pain meds onto people whose pain is really from their economic and social handicaps.

Opioids are not the right treatment for that kind of pain, but psychedelics are.

Commonly, what begins as a legally prescribed medical treatment using opioids ends up with the person seeking the drugs on the black market. That may lead them to fentanyl, a synthetic opiate that is lethal in three milligram doses and extremely dangerous but relatively cheap. Easily made and mixed with other drugs, it is smuggled into the U.S., mostly from Mexico and China, where huge quantities are made in illicit laboratories. We can demand that China close the labs, but that would be unlikely to work in a trade war environment if there are people willing to pay for the drug. In a similar situation our colonist ancestors were involved in a land dispute with Native Americans and did not hesitate to provide them blankets infested with smallpox that were just as lethal to them as fentanyl is to us. And if asking worked in the drug war, Mexico would have figured out long ago how to stop the flow of drugs at their border.

And if we could learn from our past errors, we would remember that Prohibition did not work, that it fueled a black market, and that it made much of the public hostile. The OAS, as discussed earlier, is united in the position that the problem is best handled as a public health issue, that incarceration is not the answer, that organized crime lives off the profits, and that police forces of most member states are not able to effectively deal with the problem.[65] Americans are demanding the drugs, and their money is feeding the cartels and the war. The FBI did much to tame the mobs and the Mafia that grew from Prohibition, but our neighboring countries have neither the police power nor the money to counter the cartels' money and raw might. The funds our government

65 Insulza. Op. cit.

provides our Latin American neighbors to fight drugs on their own turf just increases the level of violence in those countries, so their people come here seeking asylum from the violence which we fuel in their home countries.

Ending Prohibition removed a large source of income from mob-related operations. The Mafia then moved into the drug business. As we have seen with the growth of health insurance, tactical decisions are easily thwarted and even snowball into bigger problems because they lack strategy. The way to handle black markets, then and now, is to move them into the formal economy. That does not mean placing such products on the open marketplace as we have done with alcohol. As a public health issue, these drugs are best handled by our healthcare industry. This would be a challenge for the industry, which would have to handle the heavy demand, but it would more than pay for itself. Legalization would divert money from the informal economy into a formal and taxable part of our economy. And it would be counted in the GDP.

We are talking about really big money. It is enough to make the healthcare industry's eyes grow wide at the prospects.

We could sum up this chapter as lizard-brain thinking is getting us nowhere, but we continue doing it. We make decisions based on fear and greed, and where does it get us? We've allowed our political and social leaders to feed us the fear that keeps us there for a long time. The solutions are all fundamentally the same: understand what is going on, stop thinking like reptiles. Stop reacting based on fear. Use what we know to help ourselves; and listen to those who have been left out—in the case, the OAS.

Establishing the rule of law should be a cognitive process, a reasoning process, not a tactical midbrain process, a fight-or-flight response. The Irish Constitution had an Amendment outlawing abortion. When, in the early teens, there got to be sufficient question of its appropriateness the government set up a citizen's committee to look at the issue. It was deliberately and accurately representative of the country's population, and after listening to experts from both sides they concluded with a majority opinion that the Amendment was inconsiderate and should be voted on. In 2018 the Amendment was eliminated by essentially the same majority as that seen in the committee. Such assemblies work; they were the major governing bodies of ancient Athens, except in Athens, they included only the men. In Ireland, both genders were represented. More use of such committees would empower communities to implement their own solutions to their problems. Strategy precedes tactics, but our drug laws, in particular, are tactical and imposed on us, a sure way to generate counter will, and they are not in our strategic interest.

In the next book we will explore how innovative ideas already exist that can solve the biggest and most intractable problems faced by our society and world. We have examples today of a more equal justice toward women, which followed empowering more of them. This correlates with the #MeToo movement and their increased interest in political activity, including the remaking of our national legislature to reflect the nation's demographics and the priorities of its citizens more fairly. Empowerment from the ground up is the best answer to Black Lives Matter, our drug problem, and other problems where justice is not equal.

America is ready. Its citizens have been waiting more than two centuries for it to finally live up to its promises. Many of the nation's young people are growing up with the perception that America is the bad guy, agreeing with radical Muslims that we are 'The Great Satan'. While we have our problems with managing both our political and economic power, we do try to be a good example of what a democratic nation can do to make the world a better place. The real problem is the nation's greatness, which grew from it being founded on principles that empowered individuals and communities to find better ways of organizing and governing, made it an attractive target for those whose greed and 'will to power' compelled them to move from marketplace capitalism to monopolistic oligarchy and take over the nation from within. Over the decades we have moved away from the principles of localized power that made us great, and instead, power is centralizing. Decisions are made from afar by people who have no local interests. From that point of view, people are only as good as the money that can be made from them. It is the driving force behind the dysfunction in health care, and it has spread to our government, too. And I can prescribe a treatment for it based on the principles and innovations spelled out in Commonsense Medicine. Look for it in the sequel.

This is Dr. Jones signing off for now. Please be good to yourselves and each other. Join me at CommonsenseMedicine.org as we stand together in the name of good health. And, as my mother told me as she sent me off to school every day: "Keep your nose clean." And it wasn't metaphorical; it was her remembrance of a childhood friend who had what today is called persistent rhino pus—a condition you won't have to deal with when primary airway defenses are optimal. Try to keep them that way.

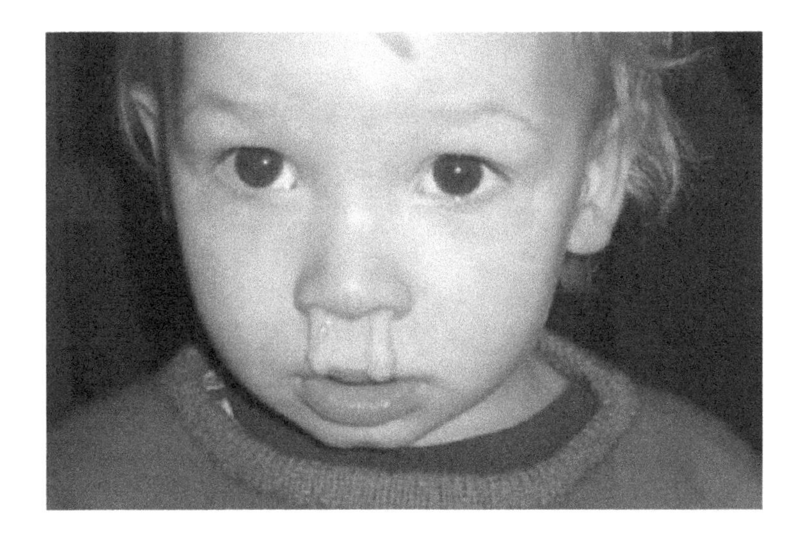

*rhino pus**

* We met this boy and his parents on a tour bus in Stehekin, Washington years ago during cold-season. He sneezed and produced this wonderfully descriptive picture. I screamed "wait wait", but Mom was too quick with her handkerchief. When I explained my need Dad was kind enough to send me the next episode along with permission to use it.

Made in USA - Kendallville, IN
55629_9781945962516
01.10.2024 1500